Student Activity Workbook

Heather Foes
Rock Valley College

Kathleen Almy
Rock Valley College

Michael Sullivan, III
Joliet Junior College

Sullivan Statistics Series
Fourth Edition

Michael Sullivan, III
Joliet Junior College

PEARSON

Boston   Columbus   Indianapolis   New York   San Francisco   Upper Saddle River
Amsterdam   Cape Town   Dubai   London   Madrid   Milan   Munich   Paris   Montreal   Toronto
Delhi   Mexico City   Sao Paulo   Sydney   Hong Kong   Seoul   Singapore   Taipei   Tokyo

The author and publisher of this book have used their best efforts in preparing this book. These efforts include the development, research, and testing of the theories and programs to determine their effectiveness. The author and publisher make no warranty of any kind, expressed or implied, with regard to these programs or the documentation contained in this book. The author and publisher shall not be liable in any event for incidental or consequential damages in connection with, or arising out of, the furnishing, performance, or use of these programs.

Reproduced by Pearson from electronic files supplied by the author.

Copyright © 2013 Pearson Education, Inc.
Publishing as Pearson, 75 Arlington Street, Boston, MA 02116.

All rights reserved. No part of this publication may be reproduced, stored in a retrieval system, or transmitted, in any form or by any means, electronic, mechanical, photocopying, recording, or otherwise, without the prior written permission of the publisher. Printed in the United States of America.

ISBN-13: 978-0-321-75912-2
ISBN-10: 0-321-75912-5

4 5 6 V036 15 14

www.pearsonhighered.com

PEARSON

Table of Contents

Activity Title (section number)

                         Page Number

Chapter 1:
Introducing Statistics Through Quotes (1.1)	3
Categorizing Student Survey Data (1.1)	5
Comparing Sampling Methods (1.4)	7
Designing an Experiment (1.6)	9

Chapter 2:
Exploring Histograms with StatCrunch (2.2)	15
Sorting Histograms by Shape (2.2)	17
Predicting Distribution Shape (2.2)	21
Constructing and Comparing Graphical Representations (2.3)	23
Recognizing and Correcting Misleading Graphs (2.4)	25

Chapter 3:
Understanding Measures of Center (3.1)	29
Comparing Statistics to Parameters (3.1 – 3.2)	31
Exploring Standard Deviation (3.2)	33
Understanding the Standard Deviation Formula (3.2)	35
Shifting and Scaling Data (3.4)	37
Matching Boxplots and Histograms (3.5)	39

Chapter 4:
Limitations of the Linear Correlation Coefficient (4.1)	43
Finding a Least-Squares Regression Line (4.2)	47
Examining the Relationship Between Arm Length and Height (4.2)	49
Minimizing the Sum of the Squared Residuals (4.2)	51
Investigating Outliers and Influential Points (4.3)	53

Chapter 5:
Demonstrating the Law of Large Numbers (5.1)	57
Finding the Probability of Getting Heads (5.1)	59
Calculating the Probability of Winning the Lottery (5.5)	61
Exploring the Duplicate Birthday Problem (5.6)	63

Chapter 6:
Finding the Expected Value of a Game (6.1)	67
Exploring a Binomial Distribution from Multiple Perspectives (6.2)	69
Using Binomial Probabilities in Baseball (6.2)	73

Chapter 7:
    Constructing Probability Distributions Involving Dice (7.1)   77
    Grading on a Curve (7.3)   79
    Modeling with the Normal Distribution (7.3)   81
    Playing Plinko (7.5)   83
    Analyzing Standardized Test Scores (7.5)   85

Chapter 8:
    Creating a Sampling Distribution for the Mean (8.1)   89
    Analyzing the Variability in Sample Means (8.1)   91
    Simulating IQ Scores (8.1)   93
    Sampling from Normal and Non-normal Populations (8.1)   95
    Creating a Sampling Distribution for a Proportion (8.2)   97
    Describing the Distribution of the Sample Proportion (8.2)   99

Chapter 9:
    Exploring the Effects of Confidence Level and Sample Size (9.1)   103
    Constructing a Confidence Interval with M&M's (9.1)   105
    Exploring the Effects of Confidence Level, Sample Size, and Shape (9.2)   107
    Constructing a Confidence Interval from a Non-Normal Distribution (9.2)   109
    Constructing a Confidence Interval for Die Rolls (9.2)   111
    Finding a Bootstrap Confidence Interval (9.5)   113

Chapter 10:
    Interpreting *P*-Values (10.2)   117
    Testing a Claim with Skittles I (10.2)   119
    Understanding Type I Error Rates I (10.2)   121
    Testing Cola Preferences (10.2)   123
    Analyzing a New Math Program (10.2)   125
    Testing a Claim with Skittles II (10.3)   127
    Understanding Type I Error Rates II (10.3)   129
    Using Bootstrapping to Test a Claim (10.5)   131
    Computing the Power of a Test (10.6)   135

Chapter 11:
    Making an Inference about Two Proportions (11.1)   139
    Analyzing Rates of Drug Side Effects (11.1)   141
    Comparing Arm Span and Height (11.2)   143
    Using Randomization Test for Independent Means (11.3)   145
    Differentiating Between Practical and Statistical Significance (11.3)   147

Chapter 12:
    Performing a Goodness-of-Fit Test (12.1)   151
    Testing for Homogeneity of Proportions (12.2)   153

Chapter 13:
    Designing a Randomized Complete Block Design (13.3)   159
    Performing a Two-Way ANOVA (13.4)   161

Chapter 14:
    Testing the Significance of a Regression Model (14.1)   165
    Using a Randomization Test for Correlation (14.1)   167

The subject of statistics, like many things in life, is best learned through active participation. These activities have been designed to improve your study of statistics by helping you become a more active participant. Whether you are collecting data, discussing concepts with fellow students, or manipulating an applet, you will need to be actively involved to get the most out of these activities. The more you engage with the material, the more you will understand the concepts and the more able you will be to apply them in real life.

Even if your instructor does not assign these activities or use them in class, there are many that you can work through on your own to increase your understanding of statistical concepts.

Chapter 1

Data Collection

Introducing Statistics Through Quotes

Instructions: Choose a word from this word bank to complete each quote about statistics.

numbers	lottery	experiment	truth	uncertainty
illumination	certain	accuracy	assumptions	approximate
averages	conclusions	suggestive	opinion	statistic
statistically	probabilities	languages	make	measure

1. Statistics may be defined as a "body of methods for making wise decisions in the face of _____." ~W.A. Wallis

2. Without data, all you are is just another person with an _____.

3. The most misleading _____ are the ones you don't even know you're making.

4. Torture _____, and they'll confess to anything. ~Gregg Easterbrook

5. When you can _____ what you are speaking about and express it in numbers, you know something about it; but when you cannot measure it, when you cannot express it in numbers, your knowledge is of the meager and unsatisfactory kind. ~Lord Kelvin (British physicist)

6. A(n) _____ answer to the right question is worth a great deal more than a precise answer to the wrong question. ~The first golden rule of mathematics, sometimes attributed to John Tukey

7. A knowledge of statistics is like a knowledge of foreign _____ or of algebra; it may prove of use at any time under any circumstances. ~A. L. Bowley

8. Statistics means never having to say you're _____. ~Unknown

9. There are two kinds of statistics, the kind you look up and the kind you _____ up. ~Rex Stout

10. Statistics are like bikinis. What they reveal is _____, but what they conceal is vital. ~Aaron Levenstein

11. I abhor _____. I like the individual case. A man may have six meals one day and none the next, making an average of three meals per day, but that is not a good way to live. ~Louis D. Brandeis

12. The death of one man is a tragedy. The death of millions is a _____. ~Joseph Stalin

4   Introducing Statistics Through Quotes

13. I could prove God _____. Take the human body alone - the chances that all the functions of an individual would just happen is a statistical monstrosity. ~George Gallup

14. The theory of _____ is at bottom nothing but common sense reduced to calculus. ~Laplace

15. _____: A tax on people who are bad at math. ~Unknown

16. Hysteria sells -- and _____ takes time, which could make the news stale by the time the statisticians check it out. ~Thomas Sowell

17. I can prove anything by statistics except the _____. ~George Canning

18. He uses statistics as a drunken man uses lampposts - for support rather than for _____. ~Andrew Lang

19. Statistics: The only science that enables different experts using the same figures to draw different _____. ~Evan Esar

20. All life is a(n) _____. The more experiments you make, the better. ~Ralph Waldo Emerson

Comparing Sampling Methods

> In this activity, you will explore how different sampling techniques affect sample statistics.
>
> Consider the following question recently asked by the Gallup Organization:
>
> "In general, are you satisfied or dissatisfied with the way things are going in the country?"
>
> You will collect responses to this question using several different sampling techniques and compare the results.

Simple Random Sample

1. Number the students in your class from 1 to N, where N is the number of students in the class.

2. Choose a sample size, n. _____

3. Use a random number generator to choose n students from the class to answer the question.

4. Record the number of satisfied responses and the number of dissatisfied responses.
   Number satisfied:_____   Percent satisfied: _____
   Number dissatisfied: _____   Percent dissatisfied: _____

Stratified Sample

1. Divide the students in the class by gender. Treat each gender as a stratum.

2. Choose a sample size, n, and decide how many students to sample from each stratum.
   Sample size: _____
   Number from female stratum: _____
   Number from male stratum: _____

3. Use a random number generator to choose the sample in each stratum.

4. Record the number of satisfied responses and the number of dissatisfied responses.
   Females: Number satisfied:_____   Percent satisfied: _____
               Number dissatisfied: _____   Percent dissatisfied: _____

   Males: Number satisfied:_____   Percent satisfied: _____
               Number dissatisfied: _____   Percent dissatisfied: _____

Cluster Sample

1. Treat each row of desks in the room as a cluster.

2. Decide how many clusters to sample.

3. Use a random number generator to obtain a simple random sample of clusters. Have each student in the selected clusters answer the question.

8   Comparing Sampling Methods

4. Record the number of satisfied responses and the number of dissatisfied responses.
   Number satisfied: _____   Percent satisfied: _____
   Number dissatisfied: _____   Percent dissatisfied: _____

Systematic Sample

1. Number the students in your class from 1 to N, where N is the number of students in the class.

2. Choose an appropriate value of k so you can survey every k^{th} student in the class. This can be done by computing the ratio N/n and then rounding down to the nearest integer.
   Value of k: _____

3. Choose a starting student using a number between 1 and k.
   Starting number, p: _____

4. Obtain the systematic sample and record the number of satisfied responses and the number of dissatisfied responses.
   Number satisfied: _____   Percent satisfied: _____
   Number dissatisfied: _____   Percent dissatisfied: _____

Conclusions

1. How did the percent satisfied compare for the various sampling techniques? Explain.

2. Which sampling technique do you think was *most* appropriate for this situation? Explain.

3. Which sampling technique do you think was *least* appropriate for this situation? Explain.

4. What are two advantages of stratified sampling over simple random sampling?

5. What are two practical advantages of doing a systematic sampling technique?

Designing an Experiment

> Suppose you are a toy designer who wishes to develop a new paper airplane. The design for the plane is complete, and it must be made on a 9-inch square piece of paper. However, it is unclear whether the plane should be made with newspaper-weight paper, brown-bag paper, or 24-pound printer paper. Your company wants to choose a paper that will maximize the distance that the plane will fly. This activity will guide you in designing and conducting an experiment to decide which paper should be used.

Step 1: Identify the problem to be solved by writing the research question. Also identify the response variable.

Research question:

Response variable:

Step 2: List at least three factors you think will affect the value of the response variable.

1.
2.
3.
4.
5.

Step 3: Determine the number of experimental units (decide on the number of planes to make). Decide how many will be made and by whom. Consider how long it will take you to make each plane and also how long it will take to fly that many planes.

Step 4: Determine the level of each factor. Consider the list of factors from Step 2 and decide if each should be controlled, fixed, or manipulated. Place an X in the appropriate box for each factor.

Factor	Controlled?	Fixed?	Manipulated?

10 Designing an Experiment

Step 5: Conduct the experiment.

    a.   Make the planes. Go to a website that suggests a plane design or use one of your own.

    b.   As a class, conduct your experiment. First, make sure you have considered how you will measure the flights, which unit of measurement you will use, and which measuring device you will use.

           Unit of measure: _____

           Measuring device: _____

    c.   Prepare to collect the data. Decide what you will need to record and set up a table to record your raw data.

    d.   Conduct the experiment and record the results.

Step 6: Test the claim.

    a.   Analyze your table and compare lengths of flights against the type of paper used.

          Which paper seemed to yield the longest flight? _____

    b.   Write a statement for the toy company officially recommending one of the papers.

Discuss any problems encountered during the experiment. Is there anything you would change in the design? In particular, are there any factors you discovered after the onset of the experiment that you believe should have been controlled that were not controlled?

What role, if any, did randomization play in your experiment?

Chapter 2

Organizing and Summarizing Data

Exploring Histograms with StatCrunch

> In this activity, you will use StatCrunch to make several different histograms and explore how the starting point and bin width affect the look of the histogram. You will make a histogram that you think represents the data set well. You will also construct one histogram that you think is not a good representation and explain why it is not.

1. Choose a data set.

    a. In StatCrunch, open the data set that contains your class data from the Student Data Survey in chapter one or open another data set as directed by your instructor.

    b. Choose a variable from the data set that you would like to explore.

    Variable selected: _____

2. Use the *Histogram with Sliders Applet* to make a histogram for your variable.

    a. This applet can be found by selecting ***StatCrunch →Applets →Histogram with sliders***.

    b. Choose your variable when asked to select a column.

    c. Select *Create Interactive*!

3. Explore the effect of changing the starting point for the histogram.

    a. The *Starting point* on the histogram applet is the _____ for the frequency distribution.

    b. Place your mouse cursor on the *Starting point* slider. Change the starting point while watching the effect this has on the histogram.

    c. By changing the starting point, make at least two histograms that have very different shapes. Print each histogram and attach them here in the space provided. Comment on the shape that each histogram implies for the distribution of your variable.

16   Exploring Histograms with StatCrunch

4. Explore the effect of changing the bin width for the histogram.

   a. The ***Bin width*** on the histogram applet is the _____ for the frequency distribution.

   b. Place your mouse cursor on the ***Bin width*** slider. Change the bin width while watching the effect this has on the histogram.

   c. By changing the bin width, make at least two histograms that have very different shapes. Print each histogram and attach them here in the space provided. Comment on the shape that each histogram implies for the distribution of your variable.

5. Make a histogram that represents the data well and a histogram that does not.

   a. By changing the starting point and the bin width, make a histogram that you think represents the distribution of your selected variable well. Print the histogram and attach it in the space provided. Comment on the shape of the distribution of your selected variable.

   b. By changing the starting point and the bin width, make a histogram that you think does not represent the distribution of your selected variable well. Print the histogram and attach it in the space provided. Comment on why this is not a good histogram for your selected variable.

Sorting Histograms by Shape

Histograms can reveal important information about the shape of a distribution for a variable. Common distribution shapes are bell-shaped, uniform, skewed right, and skewed left.

Study the histograms shown on the next pages and classify them according to their shape. In the table below, list the categories and the graphs that you included in each category. You should feel free to create categories other than those mentioned here.

Please remember that identifying the shape of a distribution from a histogram is subjective. Often a histogram will not perfectly match any shape, and you will have to use your best judgment to classify it.

Category Name	Category Description	List of Graphs Included

Copyright © 2013 Pearson Education, Inc.

18   Sorting Histograms by Shape

1.
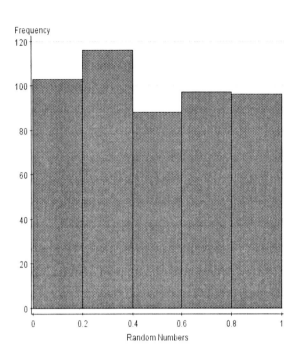

2.
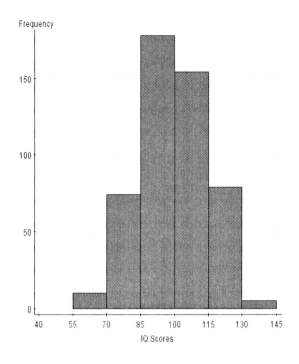

3.
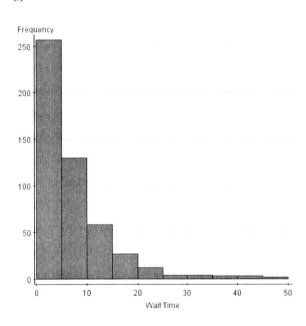

4.
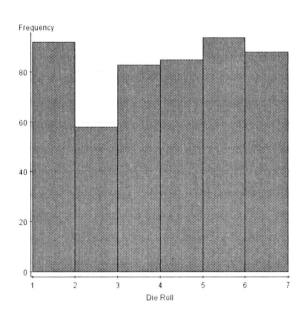

5.

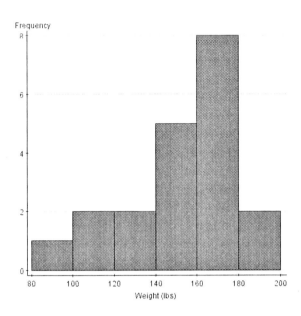

6.

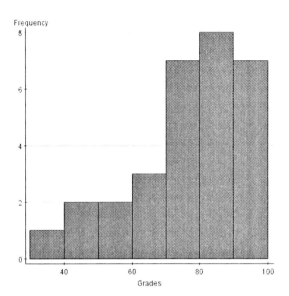

7.

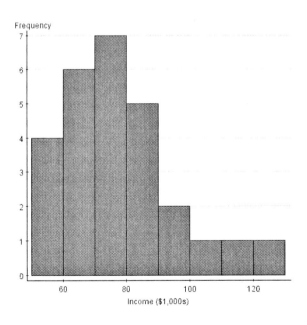

8.

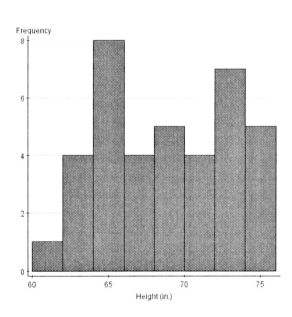

20   Sorting Histograms by Shape

9.

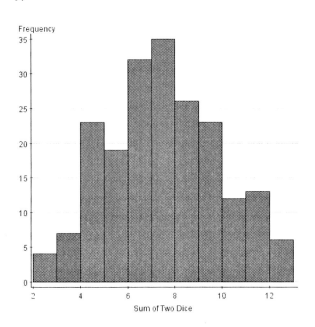

10.

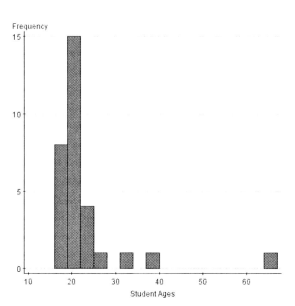

11.

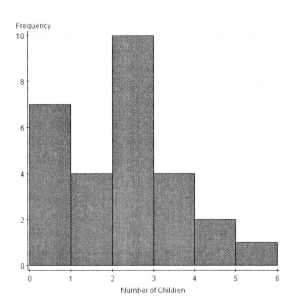

12.

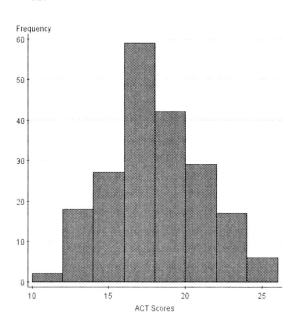

Predicting Distribution Shape

> Sometimes the shape of a distribution can be predicted based on the context of the data. This activity will give you an opportunity to practice predicting the shape of the distribution for data collected from your class. You will then construct histograms, identify the shape of the distribution, and judge how accurate your prediction was.

1. Predict the shape of a distribution.

    For each of the following, identify the shape you expect for the distribution. Choose between uniform, bell-shaped, skewed right, and skewed left. Write a sentence to explain why you expect that particular shape for the data.

    a. Random integers between 0 and 20, inclusive, generated by a calculator or statistical software

    _____

    b. Resting pulse rates for your classmates

    _____

    c. Size of household for your classmates

    _____

    d. Ages of your classmates

    _____

2. Collect the data.

    Follow your teacher's instructions to collect data for each of the variables mentioned in question 1. Store the data in your calculator or statistical software so you can make histograms from it.

3. Make histograms.

    Construct a histogram for each data set you collected. If you are making histograms by hand, draw them on the next page. If you are using technology, print them and attach them on the next page.

22 Predicting Distribution Shape

4. Analyze the shapes of the histograms.

   For each data set, identify the shape of the data set based on the histogram you constructed. Then compare this shape to the shape you predicted. Comment on any differences between what the shape expected and the shape you found.

   a. Random integers between 0 and 20

      Predicted shape: Actual shape:

      Comment:

   b. Resting pulse rates

      Predicted shape: Actual shape:

      Comment:

   c. Size of household

      Predicted shape: Actual shape:

      Comment:

   c. Ages

      Predicted shape: Actual shape:

      Comment:

Constructing and Comparing Graphical Representations

> This activity will give you an opportunity to practice constructing graphs, either by hand or using StatCrunch. Before making a graph, you will need to consider the advantages and disadvantages of different graphical representations and choose ones that will represent the data well. After constructing your graphs, you will talk to other students and share the graphs you made.

1. In StatCrunch, open the data set that contains your class data from the Student Data Survey in chapter one or open another data set as directed by your instructor.

    Your instructor will assign you a variable to analyze from this data set.

    Variable name: _____

    Qualitative or quantitative: _____

    Level of measurement: _____

2. Make at least two different graphs to represent your data. Draw the graph if you are making it by hand or print the graph from StatCrunch and attach it. For each graph, explain why you chose that particular type of graph and explain what it tells you about the data.

    Graph #1:

    Graph #2:

24   Constructing and Comparing Graphical Representations

3. Form a group with other students who worked on the same variable as you did. Show each other the graphs you made and discuss why you chose the particular graphs that you did.

   Which graph was selected most often for your variable? Why?

   Were there any graphs that were never made for your variable? If so, why were those graphs not selected?

   Summarize what you learned about your variable from the graphs that were created by your group.

Recognizing and Correcting Misleading Graphs

Unfortunately, there are many misleading and deceptive graphs published in newspapers, magazines, and in online sources. This activity will involve finding a graph that you think is misleading or deceptive and recreating the graph to give a more accurate impression.

1. Find a graph that you think is misleading or deceptive and attach it here. You might consider newspapers, magazines, or online sources.

2. Discuss the difference between a graph that misleads and a graph that deceives (see section 2.4 in the textbook). Do you think the graph you found is misleading or deceptive? Explain why.

26 Recognizing and Correcting Misleading Graphs

3. Recreate the graph to address the issue that made it misleading or deceptive.

Chapter 3

Numerically Summarizing Data

Understanding Measures of Center

> In this activity you will use an applet to investigate the mean and median as measures of center and explore the resistance of these measures.

Load the *Mean Versus Median Applet* that is located on the CD that accompanies the text or from the Multimedia Library in MyStatLab.

1. Change the lower limit to 0 and the upper limit to 10. Click update.

    a. Create a data set of ten observations such that the mean and median are both roughly equal to 2.

    b. Now add a new observation near 9. How does this new value affect the mean? The median?

2. Change the upper limit to 25. Click update, but do not Trash your observations.

    a. Drag the single observation that is near 9 so that it is near 24 instead.

    b. How does this new value affect the mean? The median?

3. Trash your observations. Change the upper limit to 50 and click update.

    a. Create a data set of ten observations such that the mean and median are both roughly equal to 40 and every observation is in the range 36 - 44.

    b. Now add a new observation near 30. How does this new value affect the mean? The median?

    c. Drag the single observation that is near 30 so that it is near 0 instead. What happens to the mean? What happens to the median?

30 Understanding Measures of Center

4. Write a paragraph that summarizes what you have learned in this activity about the mean and median. Be sure to include a discussion of the concept of resistance.

5. Trash your observations. Change the upper limit to 10 and click update.

    a. Create a data set of at least ten observations such that the mean equals the median. Explain how you accomplished this.

    b. Create a data set of at least ten observations such that the mean is greater than the median. Explain how you accomplished this.

    c. Create a data set of at least ten observations such that the mean is less than the median. Explain how you accomplished this.

    d. Create a data set that is skewed left, with at least 50 observations. Describe the relationship between the mean and the median.

    e. Create a data set that is skewed right, with at least 50 observations. Describe the relationship between the mean and the median.

    f. Will the relationships between the mean and median that you observed in parts d and e always hold? Can you create a data set for which they do not hold?

Comparing Statistics to Parameters

> You will begin this activity by collecting data from the population of students in your class. Once you have collected the data, you will draw samples from the population and compare the sample mean and standard deviation to the population mean and standard deviation for your particular data.

Treat the students in your class as a population. All the students in the class should determine their value of a variable. You might collect values of pulse rates, heights, travel time to school, number of texts sent/received yesterday or some other variable as directed by your instructor.

1. Number each student in the class from 1 to N, where N is the number of students in the class. Write each student's corresponding value of the variable next to their number. Add more blanks as needed.

   1. _____ 2. _____ 3. _____ 4. _____

   5. _____ 6. _____ 7. _____ 8. _____

   9. _____ 10. _____ 11. _____ 12. _____

   13. _____ 14. _____ 15. _____ 16. _____

   17. _____ 18. _____ 19. _____ 20. _____

   21. _____ 22. _____ 23. _____ 24. _____

   25. _____ 26. _____ 27. _____ 28. _____

   29. _____ 30. _____ 31. _____ 32. _____

   33. _____ 34. _____ 35. _____ 36. _____

   37. _____ 38. _____ 39. _____ 40. _____

2. Determine the population mean, variance, and range of the variable.

3. Obtain a simple random sample of $n = 5$ students and record the data here.

   Determine the sample mean: _____ Does the sample mean equal the population mean?

   Determine the sample variance: _____ Does the sample variance equal the population variance?

   Determine the sample range: _____ Does the sample range equal the population range?

36   Understanding the Standard Deviation Formula

3. Find the variance and standard deviation for each sample (using the formula that involves division by $n-1$). If you are using technology for the computation, the standard deviation would be labeled s not sigma.

   List the values in the 3rd and 4th columns of the table.

4. Find the variance and standard deviation for each sample by treating the sample as a population (using the formula that involves division by n). If you are using technology for the computation, the standard deviation would be labeled sigma, not s.

   List the values in the 5th and 6th columns of the table.

5. Find the mean for each of the 3rd, 4th, 5th, and 6th columns. List them in the last row of the table.

6. If the mean of a statistic (over all possible samples of the same size) is equal to the population parameter, the statistic is said to be an unbiased estimator of that parameter.

   Which method (dividing by $n-1$ or dividing by n or neither) resulted in an unbiased estimate of the population variance?

   Which method (dividing by $n-1$ or dividing by n or neither) resulted in an unbiased estimate of the population standard deviation?

32 Comparing Statistics to Parameters

4. Obtain a second simple random sample of $n = 5$ students and record the data here.

   Determine the sample mean. _____ Does the sample mean equal the population mean?

   Determine the sample variance: _____ Does the sample variance equal the population variance?

   Determine the sample range: _____ Does the sample range equal the population range?

5. Obtain a third simple random sample of $n = 5$ students and record the data here.

   Determine the sample mean: _____ Does the sample mean equal the population mean?

   Determine the sample variance: _____ Does the sample variance equal the population variance?

   Determine the sample range: _____ Does the sample range equal the population range?

6. Are the three sample means the same? Are the three sample variances the same? Why or why not?

7. The sample mean and variance are said to be *unbiased estimators* of the population mean and variance. Write a paragraph to explain what this term means.

8. The sample range is not an unbiased estimator of the population range. Write a paragraph to explain what this means.

Exploring Standard Deviation

> In this activity you will use an applet to investigate the standard deviation of a data set. You will investigate how to make the standard deviation smaller or larger as well as the how the standard deviation is affected by extreme values.

Load the **Standard Deviation Applet** that is located on the CD that accompanies the text or from the Multimedia Library in MyStatLab.

1. Change the lower limit to 0 and the upper limit to 10. Click update.

    a. Create a data set with ten observations so that the standard deviation is 0.5. Record a graph of your data.

    b. By changing your previous data set or creating a new one, construct a data set so that the standard deviation is 1.0. Record a graph of your data.

    c. By changing your previous data set or creating a new one, construct a data set so that the standard deviation is 2.0. Record a graph of your data.

    d. Describe each of the three data sets you created in terms of their spread.

2. Create a data set with ten observations so that the standard deviation is 0.5. Record a graph of your data. Grab one of the points with your mouse cursor and move it away from the rest of the data. How does the standard deviation change? Is the standard deviation resistant?

34   Exploring Standard Deviation

Trash your observations. The lower limit should still be set to 0 and the upper limit to 10.

   a.   Create a data set with ten observations so that the standard deviation is as large as possible. What is the largest standard deviation you are able to achieve? Record a graph of your data.

   b.   Create a data set with ten observations so that the standard deviation is as small as possible. What is the smallest standard deviation you are able to achieve? Record a graph of your data.

   c.   Is it possible to create a data set with a standard deviation of 0? If so, what does the distribution look like?

Understanding the Standard Deviation Formula: Why Divide by $n - 1$?

> This activity will give you an opportunity to better understand the formula for the sample standard deviation and why it involves division by $n - 1$ instead of n. You will also get some practice finding sample and population standard deviations and variances, whether by hand or using technology.

Consider a family of four people, aged 8, 10, 33, and 37, as a population.

1. Calculate the population variance and standard deviation for the ages.

   Variance: _____   Standard deviation: _____

2. Record each of the possible 16 samples of size two from this population in the second column of the chart. Assume that random samples of size two are drawn with replacement from this population and their ages are recorded. Instructions for completing the rest of the table continue on the next page.

Sample	Sample Values	SD (treated as a sample)	VAR (treated as a sample)	SD (treated as a population)	VAR (treated as a population)
1					
2					
3					
4					
5					
6					
7					
8					
9					
10					
11					
12					
13					
14					
15					
16					
		Mean =	Mean =	Mean =	Mean =

Copyright © 2013 Pearson Education, Inc.

Scaling and Shifting Data

> This activity will help you consider how transforming data affects the mean, standard deviation, and shape of a distribution.

1.
   a. Use StatCrunch to generate 40 exam scores from a normal distribution with a mean of 75 and a standard deviation of 10. Name the column "Exam Scores."

   b. Calculate the mean and standard deviation for your exam scores.
      Mean = _____ Standard deviation = _____

   c. Choose one of the exam scores and calculate its z-score.

   d. Create a histogram for the exam scores and sketch it here.

2. Suppose the teacher plans to adjust the scores by adding ten points to each student's score.

   a. Transform the exam score data by adding ten points to each score. Create a new column and name it "Exam Scores + 10."

   b. Calculate the mean and standard deviation for the adjusted exam scores.

      Mean = _____ Standard deviation = _____

   c. Calculate the z-score for the same student whose score you used in question 1.

   d. Create a histogram for the exam scores and sketch it here.

38   Scaling and Shifting Data

e. Describe how adding ten points to the exam scores affected the mean, standard deviation, and the shape of the distribution. How did the adjustment affect the student's z-score?

3. Suppose the teacher plans to adjust the scores by giving each student a score 20% higher than his/her original score.

   a. Transform the exam score data by multiplying each score by 1.2. Create a new column and name it "Exam Scores x 1.20."

   b. Calculate the mean and standard deviation for the adjusted exam scores.

   Mean = _____ Standard deviation = _____

   c. Calculate the z-score for the same student whose score you used in question 1.

   d. Create a histogram for the exam scores and sketch it here.

   e. Describe how adjusting the scores by multiplying by 1.20 affected the mean, standard deviation, and the shape of the distribution. How did the adjustment affect the student's z-score?

Matching Boxplots and Histograms

> Match each histogram on this page with the appropriate boxplot on the next page. Place the letter of the correct boxplot in the blank next to each histogram.

1. _____

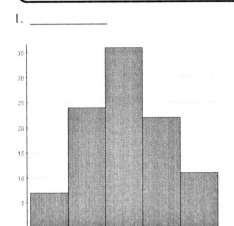

2. _____

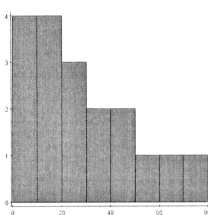

3. _____

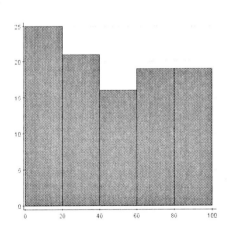

4. _____

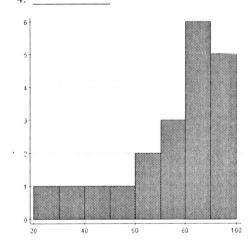

5. _____

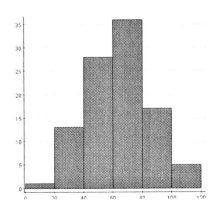

6. Write a paragraph to explain how you matched the histograms and boxplots. Which characteristics of the distributions were you able to see in the graphs that helped you match them?

40 Matching Boxplots and Histograms

A.

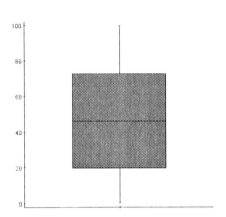

B.

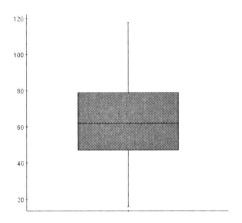

C.

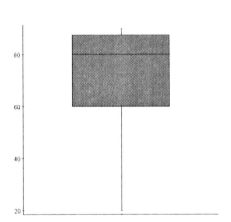

D.

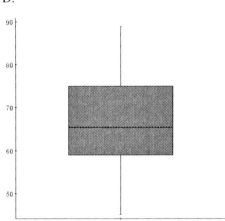

E.

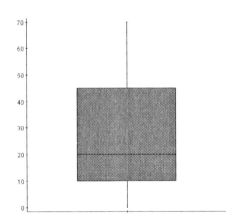

Chapter 4

Describing the Relation between Two Variables

Limitations of the Linear Correlation Coefficient

> In this activity you will use an applet to create scatter diagrams and calculate the linear correlation coefficient. You will have the opportunity to observe important properties and limitations of the linear correlation coefficient.

Load the *Correlation by Eye Applet* that is located on the CD that accompanies the text or from the Multimedia Library in MyStatLab. Remove any points that currently exist on the scatter diagram by clicking on the Trash can.

1. a. In the lower-left corner of the applet, add ten points that are positively associated so that the linear correlation coefficient of the data is approximately 0.8. Click "show r" to show the value of the linear correlation coefficient. Draw the scatter diagram and record the correlation coefficient below.

   b. Add another point in the upper-right corner of the applet that roughly lines up with the ten other points you have in the lower-left. Draw the scatter diagram below and comment on how the linear correlation coefficient changes.

   c. Drag the point in the upper-right corner straight down. Draw the scatter diagram below and comment on how the linear correlation coefficient changes. What property of the linear correlation coefficient does this illustrate?

44   Limitations of the Linear Correlation Coefficient

2. Remove the points from the first exercise by clicking on the Trash can.

    a. Create a scatter diagram that consists of approximately ten points that for an upside-down U pattern on the applet. Click "show r."

    Draw the scatter diagram and record the value of the linear correlation coefficient.

    Would you say that no relation exists between the two variables?

    What does this activity say about what the linear correlation coefficient measures?

3. Remove the points from the last exercise by clicking on the Trash can.

    a. Create a scatter diagram that consists of approximately ten points such that the linear correlation coefficient is approximately 0.75. Draw the scatter diagram and record the value of the linear correlation coefficient.

    b. Clear the applet by clicking on the Trash can. Plot approximately six points vertically on top of each other on the left side of the applet. Add a seventh point to the right side of the applet. Move the point around until the linear correlation coefficient is close to 0.75. Draw the scatter diagram and record the value of the linear correlation coefficient. On your drawing, highlight the seventh point.

c. Clear the applet. Plot approximately seven points in a U-shaped curve near the lower left corner of the applet. Add an eighth point and move the point around until the linear correlation coefficient is close to 0.75. Draw the scatter diagram and record the value of the linear correlation coefficient. On your drawing, highlight the eighth point.

d. Write a paragraph that explains the pitfalls of relying exclusively on the linear correlation coefficient to suggest the strength of the linear relationship between two quantitative variables.

Limitations of the Linear Correlation Coefficient

Finding a Least-Squares Regression Line

> This activity involves measuring stacks of paper to create a model that will predict the thickness of a stack of paper based on the number of pages. You will have the opportunity to compare your model to others in the class and to use your model to make predictions.

1. Use a ruler to measure the thickness of a single page of the textbook. Explain why you chose the particular unit that you used.

2. Grouping pages (one page is one sheet of paper), measure to complete the following table:

No. of Pages	25	50	75	150	200	225
Thickness						

3. Compute the least-squares regression line for your data in the table.

4. Compare your data and your regression line to other students' work. Did everyone get the same measurements and model? Explain why or why not.

5. Use your model to estimate the thickness of a single page of the textbook by letting $x = 1$. Is the value you obtained reasonable? Explain.

6. Use your model to estimate the thickness of a single page of the textbook by interpreting the slope. Is the value you obtained reasonable? Explain.

7. What, if anything, could you do to improve the model?

Copyright © 2013 Pearson Education, Inc.

Examining the Relationship Between Arm Length and Height

> This activity uses arm length and height data collected from your class to determine a linear model. You will create a scatter diagram, find a regression equation, and then use the equation to make predictions.

1. Each student in your class should measure the length of their forearm and their height, both as accurately as possible in centimeters. Collect the paired data (forearm length, height) and record it here.

   _____   _____   _____
   _____   _____   _____
   _____   _____   _____
   _____   _____   _____
   _____   _____   _____
   _____   _____   _____
   _____   _____   _____
   _____   _____   _____
   _____   _____   _____
   _____   _____   _____
   _____   _____   _____
   _____   _____   _____

2. Enter the data into lists on your calculator or StatCrunch or other software.

3. Create a scatter diagram of the paired data and record it here.

50   Finding a Least-Squares Regression Line

4. From the scatter diagram, does there appear to be a relationship between forearm length and height?

    If so, describe the relationship. Does it appear to be linear? Does it appear strong or weak? Does there appear to be a positive or negative association between the two variables?

5. Find the value of the linear correlation coefficient for this data and provide an interpretation.

6. Find the least-squares regression line that relates forearm length and height. State the equation and graph the line on the scatter diagram on the previous page.

7. Explain what the y-intercept of the regression line means in this context.

8. Explain what the slope of the regression line means in this context.

9.
    a. Use the regression equation to find the predicted height for your forearm length. _____

    b. Calculate the difference between your actual height and the height predicted by the regression equation. _____

    c. What is this difference called? _____

10. Use the regression equation to predict the height of someone with a 6-inch long forearm. Is your value reasonable? Explain why or why not.

Minimizing the Sum of the Squared Residuals

> In this activity you will use an applet to create a scatter diagram and try to determine the best-fit line by eye. You will attempt to get the sum of the squared errors for your best-fit line as low as possible and then compare your best-fit line to the regression line.

Load the ***Regression By Eye Applet*** or use the ***Regression by Eye Interactive Applet*** in StatCrunch.

1. Trash any points currently on the scatter diagram and create a new scatter diagram by plotting your own points or hitting "new data" at the bottom of the screen.

2. Use the endpoints to drag the green line until you think it is the best-fit line for the data.

    Record the sum of the squared errors for the green line. Green SSE: _____

    Record the sum of the squared errors for the regression line. Regression SSE: _____

    How close did your SSE come to the regression SSE?

    Click "show regression line" to graph the true regression line.

    How close did your line come to the least-squares regression line?

    Is it possible to fit a linear model to the data that makes the sum of the squared errors smaller than the sum of the squared errors for the least-squares regression line? Explain.

3. Trash any points currently on the scatter diagram and create a new scatter diagram by plotting your own points or hitting "new data" at the bottom of the screen.

    Repeat the process from the previous exercise. Draw the green line until you think it is the best-fit line for the data. Observe the green SSE. Keep moving the green line to get the green SSE as low as possible. Then click "show regression line" to see how close your line was to the true regression line.

52 Minimizing the Sum of the Squared Residuals

Investigating Outliers and Influential Points

> This activity will use an applet to create scatter diagrams and investigate the effect of outliers and influential points on the regression equation.

Load the **Regression by Eye Applet** that is located on the CD that accompanies the text or from the Multimedia Library in MyStatLab. Remove any points that currently exist on the scatter diagram by hitting "Trash."

1. Create a scatter diagram with approximately twelve observations clustered in the upper-right corner of the applet that have a strong negative association. Click "Show regression line" on the applet. Draw the scatter diagram created along with the regression line below.

   a. Add one more point in the bottom-right corner of the applet. The point should be near the least-squares regression line.

      How does the regression line change?

      Is this point influential?

   b. Drag the new point in the bottom-right corner straight up.

      Is the point influential now? Why?

2. Clear the applet by hitting "Trash." Create a scatter diagram with approximately twelve points with a strong negative association. Be sure the points are spread out in the applet. Click "Show regression line." Draw the scatter diagram created along with the regression line below. Also record the least-squares regression equation.

54   Minimizing the Sum of the Squared Residuals

a. Add one more point near the mean of X and the mean of Y.

Does the regression line change much?

Is this point influential?

b. Drag the new point straight down.

Does the regression line change much?

Is the point an outlier? Why or why not?

Chapter 5

Probability

Demonstrating the Law of Large Numbers

> In this activity you will use an applet to simulate flipping a fair coin and then estimate the probability of getting heads based on your results. You will explore how the probability changes as the number of trials is increased. You will then repeat this process with a weighted coin.

Load the ***Long-Run Applet*** that is located on the CD that accompanies the text or from the Multimedia Library in MyStatLab.

1. Choose the "simulating the probability of a head with a **fair** coin" applet and set $n = 10$ to simulate flipping a fair coin ten times. What is the estimated probability of a head based on these ten trials?

2. Reset the applet and simulate flipping a fair coin ten times again.

   a. What is the estimated probability of a head based on these ten trials?

   b. Compare the result to the probability obtained in the first ten trials.

3. Reset the applet and simulate flipping a fair coin one thousand times. Uncheck the "Animate" box to do this more quickly.

   a. What is the estimated probability of a head based on these one thousand trials?

   b. Describe what happens to the graph as the number of trials of the probability experiment increases.

   c. What law does this illustrate?

58   Demonstrating the Law of Large Numbers

4. Reset the applet and simulate flipping a fair coin one thousand times again.

    a. What is the estimated probability of a head based on these one thousand trials?

    b. Compare the result to the probability obtained in the first one thousand trials.

5. Choose the "simulating the probability of a head with an **unfair** coin P(H) = 0.2" applet and simulate flipping this coin ten times.

    a. Record the ten outcomes.

    b. If you did not know the coin was unfair, would you consider your ten trials sufficient evidence to conclude that the coin was not weighted fairly? Why or why not?

    c. Compare your results with those of your classmates. How many simulations did not yield "sufficient" evidence?

Finding the Probability of Getting Heads

> People tend to assume that a fair penny has a 50% chance of landing on heads and a 50% chance of landing on tails. Do you think it matters how the penny is flipped or tossed? This activity will give you the opportunity to generate data with several different methods of flipping a penny. For each method, you will estimate the probability of getting heads and then compare your results.

1. Toss a penny in the air, catch it with one hand, and turn it over on your other hand. Record which side (heads or tails) is facing up when it lands. Complete a total of 25 trials with this method, recording your result. Combine your results with the results from 3 other students to obtain a total of 100 trials.

   From your experimental data, estimate the probability of getting heads with this method.

2. Balance a penny on its edge on your desk or table, slam your hand next to it to make the penny fall. Record which side (heads or tails) is facing up when it lands. Complete a total of 25 trials with this method, recording your result. Combine your results with the results from 3 other students to obtain a total of 100 trials.

   From your experimental data, estimate the probability of getting heads with this method.

3. Spin a penny on its edge and let it fall on its own. Record which side (heads or tails) is facing up when it lands. Complete a total of 100 trials with this method, recording your results.

   From your experimental data, estimate the probability of getting heads with this method.

Copyright © 2013 Pearson Education, Inc.

60   Finding the Probability of Getting Heads

4. Compare your probabilities from the first three questions. Were they significantly different? Did any of the results surprise you?

5.
   a. Collect the probabilities of getting heads with the first method from your classmates. Illustrate these probabilities with a dot plot. Comment on the plot.

   b. Collect the probabilities of getting heads with the second method from your classmates. Illustrate these probabilities with a dot plot. Comment on the plot.

   c. Collect the probabilities of getting heads with the third method from your classmates. Illustrate these probabilities with a dot plot. Comment on the plot.

   d. Comment on any differences you notice in the three plots.

Calculating the Probability of Winning the Lottery

> In this activity, you will use your knowledge of counting techniques to confirm the chances of winning a particular lottery game.

The following information is available on the Illinois Lottery's website.

The **LITTLE LOTTO GAME** awards prizes to ticket holders matching two (2), three (3), four (4) or five (5) numbers from one (01) through thirty-nine (39) with the five (5) Little Lotto Game Winning Numbers randomly selected during the scheduled drawing.

WAYS TO WIN

Prize	Chances	AVG Prize
First Prize Match all 5 numbers in any order	1:575,757	$100,000 starting jackpot
Second Prize Match 4 of the 5 numbers in any order	1:3,387	$100
Third Prize Match 3 of the 5 numbers in any order	1:103	$10
Fourth Prize Match 2 of the 5 numbers in any order	1:10	$1
Overall Chances	1:9	

Source: "Ways to Win the Little Lotto" from www.illinoislottery.com

1. Show how to calculate each of the numbers in the "Chances" column.

2. Show how the "Overall Chances" was calculated.

62 Calculating the Probability of Winning the Lottery

3. Do some research online to find data for a different lottery game. Print the probability table from the website and attach it here.

4. Show the calculations to confirm at least one of the probabilities from your chosen lottery.

Exploring the Duplicate Birthday Problem

What is the probability of two students in your statistics class having the same birthday? Do you think that event is likely or unlikely? This activity will give you the opportunity to explore the probability of duplicate birthdays through data collection, a simulation, and a direct calculation.

1. Find out if there are any duplicate birthdays (same day but not necessarily same year) in your statistics class. One way to do this is to have all the January birthdays call out their birth date, then the February birthdays, and so on.

   How many students are in your class? _____
   Are there any duplicate birthdays? _____
   How likely do you think it is to have duplicate birthdays in a group of your size? Explain.

2. Now each student in your class will run a simulation to estimate the probability of having duplicate birthdays in a group of your size.

   a. Use a random number generator to generate a set of birthdays (numbers 1 through 365) as large as your class size. Sort the list to make checking for duplicates easier.

      Record the birthdays here. Are there any duplicates?

   b. Use the collective class data to answer the following questions.

      How many of the simulations resulted in duplicate birthdays? _____

      How many total simulations were conducted? _____

      What is your estimate for the probability of having duplicate birthdays in a group of your size, based on these simulations? _____

64   Exploring the Duplicate Birthday Problem

3. Use counting techniques to find the exact probability of having duplicate birthdays in a group of your size.

4. Compare the estimated probability from your class simulations to the exact probability that you calculated.

5. How many students would need to be in a class for there to be at least a 50% chance of having duplicate birthdays?

Chapter 6

Discrete Probability Distributions

Finding the Expected Value of a Game

> In this activity, you will play a game of chance many times and estimate the mean winnings. You will then compare your results to the expected value of the game as calculated from a probability distribution.

Consider the following game of chance. A player pays $1 and rolls a pair of fair dice. If the player rolls a sum of 2, 3, 4, 10, 11, or 12, the player loses the $1 bet. If the player rolls a sum of 5, 6, 8, or 9, there is a "push" and the player gets the $1 bet back. If the player rolls a sum of 7, the player wins $1.

1. Play the game in small groups (no money should actually change hands). Keep track of the results of the game and record the winnings for each game below (use +1 for a winner, 0 for a push, and -1 for a loser). Each group should play the game at least 25 times.

Game	1	2	3	4	5	6	7	8	9	10
Result										

Game	11	12	13	14	15	16	17	18	19	20
Result										

Game	21	22	23	24	25
Result					

2. Combine the results of all the small groups. Determine the mean winnings.

3. Construct a probability distribution that describes the game.

68   Finding the Expected Value of a Game

4. Determine the expected value of the game. Is the mean obtained in #2 close to the expected value of the game? Explain.

5. Draw a graph similar to Figure 3 in Section 6.1 which illustrates the mean earnings versus the number of games played.

Exploring a Binomial Distribution from Multiple Perspectives

> This activity will help you explore a situation involving the binomial probability distribution through a physical simulation with cards, a simulation with an applet, a mathematical model, and by using statistical software.

1. There are many estimates that exist for the percentage of the population that is left-handed. For this activity, let's assume that 10% of the population is left-handed, which is a popular estimate. We would like to build a probability model for the random variable, X, the number of left-handed people in a group of $n = 10$ individuals. One approach we could take is to simulate the situation.

    a. From a standard deck of cards, select the cards Ace through 10 of a single suit (such as hearts). Let the Ace represent the left-handed individual, and the other cards represent right-handed individuals so that the probability of obtaining a left-handed individual (an Ace) from the ten individuals is 0.1.

    Select a single card from the stack.

    If the card is an Ace, record a success, otherwise, a failure.

    Replace the card and shuffle. Repeat this nine more times.

    How many successes did you obtain? _____

    b. In part (a), we are sampling with replacement so that the outcomes of each trial are independent. Explain what this means.

    c. Explain why the scenario laid out in part (a) qualifies as a binomial experiment.

70 Exploring a Binomial Distribution from Multiple Perspectives

d. Repeat part (a) nineteen more times for a total of 20. For each repetition of the experiment, write down how many Aces you drew (the number of successes or left-handed individuals). Then use your data to start building a probability model for X, the number of left-handed people in a group of 10 individuals. Consider how many of the 20 repetitions resulted in 1 Ace, how many resulted in 2 Aces, etc.

Experiment Number	Number of Aces
1	
2	
3	
4	
5	
6	
7	
8	
9	
10	
11	
12	
13	
14	
15	
16	
17	
18	
19	
20	

x	P(x)
0	
1	
2	
3	
4	
5	
6	
7	
8	
9	
10	

e. Combine your simulation results with those of your classmates. Produce a dot plot of the combined data. Use the results from your classmates to further build the probability model.

x	P(x)
0	
1	
2	
3	
4	
5	
6	
7	
8	
9	
10	

Exploring a Binomial Distribution from Multiple Perspectives

2. Open the **Binomial Distribution Applet**. Let $n = 10$ (the ten individuals), $p = 0.1$ (probability of success), and $N = 1$ (repetitions of the binomial experiment). Press "Simulate."

   a. How many successes did you obtain in the $n = 10$ trials of the binomial experiment?

   b. Repeat the binomial experiment $N = 1000$ times. Based on these results, what is the probability of finding exactly two left-handed individuals in a group of ten individuals?

   Compare your probability to the probabilities found by your classmates. Explain why the results may differ.

3. Rather than obtaining the probability model by drawing cards or using an applet, we could develop a mathematical model. One such model for determining the number of successes x in n trials of a binomial experiment where p represents the probability of success is

$$P(x) = {}_nC_x \cdot p^x \cdot (1-p)^{n-x}$$

   a. We call this model the binomial probability distribution function (pdf). Use this model to find $P(2)$, the probability of obtaining $x = 2$ left-handed people in a group of $n = 10$ people where $p = 0.1$.

   b. How does the theoretical probability compare to the probability suggested by the simulation?

   c. What would we expect to happen as the number of simulations increases if the above binomial probability model is a good model? Why?

4. Use statistical software to build a probability model for the number of left-handed people in a group of ten people using the binomial probability distribution function. Compare the results of the simulation to the theoretical probabilities obtained from the binomial pdf.

x	P(x)
0	
1	
2	
3	
4	
5	
6	
7	
8	
9	
10	

Using Binomial Probabilities in Baseball

> This activity involves finding the probability of breaking a homerun record using simulations and the binomial probability formula. Just how unusual was it when Mark McGuire broke the homerun record in 1998?

In 1998 the baseball world was enthralled by the epic chase of Mark McGuire and Sammy Sosa to surpass the single-season homerun record of 61 set by Roger Maris in 1961.

1. a. Prior to his prolific 1998 season in which he shattered Roger Maris' single season homerun record of 61 by hitting 70 "round-trippers," Mark McGuire averaged 1 homerun every 11.9 at-bats. Assuming this rate of homerun hitting applied to the 1998 season, determine the probability McGuire hits a homerun during a randomly selected at-bat in 1998.

   b. Open the *Coin-Tossing Applet*. Enter the probability determined in part (a) in the "Probability of heads" cell. Under the number of tosses, enter 600 to represent the typical number of at-bats during the season for a starting player.

   Run a total of 20 repetitions. What does each of these 20 repetitions represent?

   Based on the graph, how many of the repetitions result in 62 or more homeruns (indicating Maris' record is broken)?

   c. There are a number of players who have averaged 1 homerun every 11.9 at-bats since Maris set his record.

   Increase the number of repetitions to 1,000 with the number of tosses at 600.

   What does each of these 1,000 repetitions represent?

   d. In the cell "As extreme as," enter ≥ 62 and select "Count." This will allow us to determine the likelihood of a player hitting 62 or more homeruns in a season to break Maris' record.

74   Using Binomial Probabilities in Baseball

2. a. Use the binomial probability formula to compute the exact probability of Mark McGuire breaking Maris' record over the course of a 20-season career assuming he averages 1 homerun every 11.9 at-bats.

    b. What does this probability change to if McGuire is able to increase his homerun rate to 1 homerun every 10.8 at-bats? This was McGuire's homerun rate in 1998.

    c. Use the binomial probability formula to compute the exact probability of any particular player (who averages 1 homerun every 11.9 at-bats) among a group of 100 players over a 10-season career (for a total of 1,000 repetitions) breaking Maris' record.

3. a. Over the past few years, the prolific homerun hitters have been averaging 1 homerun every 13 at-bats. Assuming the league has 10 prolific homerun hitters in any given season, what is the likelihood of McGuire's record being broken in the next 20 years? Assume 600 at-bats per year.

        Answer using a simulation:

        Answer using the binomial probability formula:

    b. What homerun rate would be required among the top 10 homerun hitters in order for there to be at least a 5% chance of breaking McGuire's record within the next 20 years? Assume 600 at-bats per year.

Chapter 7

The Normal Probability Distribution

Constructing Probability Distributions Involving Dice

> In this activity, you will explore different distributions that arise when dealing with dice. If you simply record the number that is rolled, what kind of distribution should you expect? If you record the sum of two dice, what kind of distribution should you expect? You will run simulations to answer these questions.

1. a. Complete the following probability distribution for the random variable X which gives the outcome when one fair die is rolled.

x	P(x)
1	
2	
3	
4	
5	
6	

   b. Is X a continuous or discrete random variable?

   How would you describe the shape of this distribution?

   What is the mean?

   c. Use StatCrunch to simulate 100 rolls of a fair die.

   Create a histogram of the results. Describe the shape of the distribution.

   Calculate the mean from the 100 rolls. How does this value compare to the mean calculated in part b?

2. a. Complete the following probability distribution for the random variable X which gives the sum when two fair dice are rolled.

x	P(x)
2	
3	
4	
5	
6	
7	
8	
9	
10	
11	
12	

   b. Is X a continuous or discrete random variable?

   How would you describe the shape of this distribution?

   What is the mean?

78 Constructing Probability Distributions Involving Dice

    c.   Use StatCrunch to simulate 100 rolls of two fair dice and find the sum for each roll.

        Create a histogram of the results. Describe the shape of the distribution.

        Calculate the mean from the 100 rolls. How does this value compare to the mean calculated in part b? How does it compare to the mean calculated in (1b)?

3.   a.   Complete a probability distribution for the random variable X which gives the average when two fair dice are rolled.

    b.   Is X a continuous or discrete random variable?

        How would you describe the shape of this distribution?

        What is the mean?

    c.   Use StatCrunch to simulate 100 rolls of two fair dice. Compute the average value of the two dice.

        Create a histogram with your results. Describe the shape of the distribution.

        Calculate the expected value of the average from the 100 rolls. How does this value compare to the mean calculated in part b? How does it compare to the mean calculated in (1b)?

4.   a.   Calculate the probability of getting a result between 3 and 4 inclusive on one roll of a fair die.

    b.   Calculate the probability of getting an average between 3 and 4 inclusive on roll of two dice.

    c.   Compare the two probabilities found in parts a and b. Which one is larger and why?

Grading on a Curve

> Suppose a professor gives a very difficult exam and wants to adjust her students' scores. This activity will allow you to explore a couple of different options for adjusted exam scores and the effects of each plan on the class scores.

1. Suppose a professor gives an exam to a class of 40 students and the scores are as follows.

35	44	46	47	47	48	49	51	53	54
55	55	57	57	57	58	59	59	59	59
60	60	60	60	60	62	62	62	64	68
69	70	72	73	73	75	75	77	82	88

   a. Find each of the following:

   Mean:

   Median:

   Standard deviation:

   Z-score for a student who originally scored 54 on the exam:

   b. Create a histogram of the scores and comment on the shape of the distribution.

2. Suppose the professor decides to adjust scores by adding 25 points to each student's score.

   a. Find each of the following after the adjustment:

   Mean:

   Median:

   Standard deviation:

   Z-score for a student who originally scored 54 on the exam:

   b. Create a histogram of the adjusted scores and compare it to the histogram created in #1.

80   Grading on a Curve

3. Suppose the professor wants the final scores to have a mean of 75 and a standard deviation of 8.

   a. To find a student's new score, first calculate their z-score and then use the z-score along with the new mean and standard deviation to find their adjusted score. Use StatCrunch to find the adjusted score for each student.

   b. Create a histogram of the adjusted scores and compare it to the histogram created in #1.

4. Suppose the professor wants the scores to be curved in a way that results in the following:

   Top 10% of scores receive an A
   Bottom 10% of scores receive an F
   Scores between the 70th and 90th percentile receive a B
   Scores between the 30th and 70th percentile receive a C
   Scores between the 10th and 30th percentile receive a D

   Find the range of scores that would qualify for each grade under this plan.

Exam Scores	Letter Grade
	A
	B
	C
	D
	F

5. Which of the three curving schemes seems the most fair to you? Explain.

Modeling with the Normal Distribution

> This activity will allow you to explore real data with a normal distribution. You will collect data, find summary statistics, and create a histogram. Finally, you will compare percentages obtained from the actual data to those found using technology.

1. What weight is indicated on each individual food package? _____

   Do you think all the packages will weigh that exact amount? Explain.

2. Weigh all the individual food packages for the class and record the data here.

   _____   _____   _____   _____
   _____   _____   _____   _____
   _____   _____   _____   _____
   _____   _____   _____   _____
   _____   _____   _____   _____
   _____   _____   _____   _____
   _____   _____   _____   _____
   _____   _____   _____   _____
   _____   _____   _____   _____
   _____   _____   _____   _____

3. Find the mean and standard deviation for the package weights.

4. Create a histogram of the package weights and comment on the shape of the distribution.

5. According to the actual data, what percent of packages had weights over _____?
   Use the normal probability commands on your calculator or software to answer this same question.
   Compare the answers you obtained from the data and technology.

6. According to the actual data, what percent of packages had weights over _____?
   Use the normal probability commands on your calculator or software to answer this same question.
   Compare the answers you obtained from the data and technology.

Playing Plinko

> This activity will allow you to investigate the probabilities involved in the game Plinko by analyzing a quincunx on which it is based.

A quincunx is a board with pegs arranged so that the first row has one peg, the second row has two pegs, the third row has three pegs and so on. Balls are dropped from the top of the board, bounce off the pegs, and land in bins below the last row of pegs.

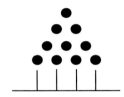

1. Draw a quincunx with 10 rows of pegs. Draw the bins at the bottom and label them 0 through 10.

2. a. A ball will land in bin 0 if it bounces to the right 0 times on its way through the quincunx. What is the probability that a ball will land in bin 0?

   b. A ball will land in bin 1 if it bounces to the right 1 time and to the left 9 times on its way through the quincunx. What is the probability that a ball will land in bin 1? Consider all the different paths that the ball could take and still land in bin 1.

   c. Complete the following table by finding the probability that a ball will land in each bin.

Bin #	0	1	2	3	4	5	6	7	8	9	10
Probability											

84 Playing Plinko

3. Graph the probability distribution that you found in the previous question and comment on its shape.

4. If 1,000 balls are dropped through the quincunx, how many would you expect to land in each bin?

Bin #	0	1	2	3	4	5	6	7	8	9	10
Expected number of balls											

5. a. Use the normal approximation to the binomial distribution to find the probability that a ball will land in bins 4 – 6.

    b. Use the normal approximation to the binomial distribution to find the probability that a ball will land in bins 0 – 1.

    c. Use the normal approximation to the binomial distribution to find the probability that a ball will land in bins 9 – 10.

Analyzing Standardized Test Scores

> This activity will allow you to investigate standardized test scores with a bell-shaped distribution and compare the probabilities obtained from the raw data to those from a normal distribution.

National ACT scores for 2005 were approximately normally distributed with a mean of 20.9 and a standard deviation of 4.9. The table shows the frequency of each score for the 1,186,251 students who took the exam that year.

ACT Composite Score	Frequency	Cumulative Percentage
36	193	99
35	1617	99
34	3729	99
33	6752	99
32	10359	99
31	15401	98
30	21725	97
29	27109	95
28	35992	93
27	44428	90
26	52811	86
25	62669	81
24	70594	76
23	77953	70
22	83849	64
21	89908	57
20	91333	49
19	89073	41
18	87315	34
17	79103	26
16	70716	20
15	59275	14
14	45796	9
13	31658	5
12	18095	2
11	6586	1
10	1594	1
9	420	1
8	138	1
7	41	1
6	12	1
5	7	1
4	0	1
3	0	1
2	0	1
1	0	1

86 Analyzing Standardized Test Scores

1. Construct a graph of the frequencies for each score on the ACT exam. Describe the shape of the distribution of scores.

2. a. Using the raw data contained in the table, what percentage of students scored between 15 and 25 (inclusive) on the test?

   b. Using the normal distribution, what percentage of students would you estimate scored between 15 and 25 (inclusive) on the test?

3. a. Using the raw data contained in the table, what percentage of students scored below 18?

   b. Using the normal distribution, what percentage of students would you estimate scored below 18?

4. a. Using the raw data contained in the table, what percentage of students scored above 28?

   b. Using the normal distribution, what percentage of students would you estimate scored above 28?

5. a. Using the raw data contained in the table, what score is in the 75th percentile?

   b. Using the normal distribution, what score would you expect to be in the 75th percentile?

6. If a college program requires a minimum score of 24 for placement, what percent of students would **not** qualify? Answer using the raw data and using the normal distribution.

Copyright © 2013 Pearson Education, Inc.

Chapter 8

Sampling Distributions

Creating a Sampling Distribution for the Mean

> In this activity, you will randomly select a small group of your classmates and then determine all the possible samples from that population. You will find the mean for each sample and then construct a sampling distribution for the mean. The goal of the activity is to better understand the shape, center, and spread of the sampling distribution for the mean.

1. Randomly select six students from the class to treat as a population. Choose a quantitative variable (such as pulse rate, age, or number of siblings) to use for this activity, and gather the data appropriately. Share the data with the class.

   Compute the mean for this population.

2. a. Work with your group to list all the samples of size $n = 2$, $n = 3$, $n = 4$, or $n = 5$, as assigned by your instructor.

   Collect the results from the other groups to complete the information for all sample sizes.

   Hint: The sample of size $n = 2$ should have $_6C_2 = 15$ possible samples.

   Samples of size $n = 2$ Samples of size $n = 3$

   Samples of size $n = 4$ Samples of size $n = 5$

   b. Compute the sample mean for each sample listed for your group's sample size in part (a).

   Record the means next to the samples.

   Collect the results from the other groups to complete the information for all sample sizes.

Creating a Sampling Distribution for the Mean

    c.   Form the probability distribution for the sample mean by making a table for each sample size and listing all the possible sample means with their probabilities.

3.   Verify for each probability distribution in (2c) that $\mu_{\bar{x}} = \mu_x$, that is, the mean of the sample means is the population mean.

4.   Draw the probability histogram for your group on the board. Compare the spread in each probability distribution based on the probability histogram. What does this result imply about the standard deviation of the sample mean?

Analyzing the Variability in Sample Means

> The purpose of this activity is to explore the amount of dispersion in values of a variable for individuals versus the variability in means.

1. Each student should determine his or her at-rest pulse rate. Be sure to agree as a class on the method for determining the pulse rate. Share the data with the class.

   Treat the students in the class as a sample of all students at the school.

   Determine the sample standard deviation of the at-rest pulse rates.

2. a. Obtain a simple random sample of $n = 5$ students from the class. Compute the mean pulse rate of the students in the sample. Repeat this five more times, for a total of six sample means.

Sample	Data Values in Sample					Sample Mean
1						
2						
3						
4						
5						
6						

   b. Compute the sample standard deviation of the mean pulse rates found in part (a).

   c. Compare the sample standard deviation of the mean pulse rates from part (b) to the sample standard deviation of the pulse rates from #1.

   Which has more dispersion, the individual pulse rates in the class, or the mean pulse rates of the five students? Why?

3. If a simple random sample of $n = 10$ students from class was used instead, would the standard deviation of the mean pulse rates be higher or lower than the standard deviation of the mean pulse rates with $n = 5$ students? Explain.

Simulating IQ Scores

> This activity will involve simulating data to represent IQ scores. After generating large samples of IQ scores, you will construct a sampling distribution for the mean and find its mean and standard deviation. You will also compare proportions calculated from the sampling distributions to those predicted by the normal distribution.

1.   a.   Use StatCrunch, MINITAB, Excel, or some other statistical software to obtain 500 random samples of size $n = 10$ from the population of IQ scores with a mean of 100 and a standard deviation of 15.

       b.   Compute the sample mean for each of the 500 samples. Provide the sample mean for Sample 1 and Sample 2 here.

                   Mean from sample 1:   _____

                   Mean from sample 2:   _____

2.   a.   What do you expect the mean and standard deviation of the sampling distribution of the mean to be?

       b.   Draw a histogram of the 500 sample means. If possible, draw the theoretical normal model on the histogram.

Simulating IQ Scores

c. Determine the mean and standard deviation of the 500 sample means. Are these values close to what was expected in part (a)?

d. What proportion of the 500 random samples resulted in a sample mean IQ greater than 108?

Based on the normal model, what proportion of random samples of size $n = 10$ would we expect to result in a sample mean greater than 108?

Is the theoretical proportion obtained from the model close to the proportion from the simulation? Explain.

Sampling from Normal and Non-Normal Populations

> In this activity, you will use an applet to simulate data from three different distributions. In each case, you will describe the center, spread, and shape of the distribution of the sample mean.

1. a. Load the **Sampling Distribution Applet** on your computer. Set the applet so that the population is bell-shaped. Decide on a mean and standard deviation and make a note of them here.

   b. Obtain 1,000 random samples of size $n = 5$, $n = 10$, and $n = 20$.

   c. Describe the distribution of the sample means based on the results of the applet. That is, describe the center (mean), spread (standard deviation), and shape of the distribution.

	$n = 5$	$n = 10$	$n = 20$
mean			
standard deviation			
shape			

   d. Compare the distributions for the various sample sizes.

   e. What role, if any, does the sample size play in the sampling distribution of the sample mean?

2. a. Set the applet so that the population is uniform.

   b. Obtain 1,000 random samples of size $n = 2$, $n = 5$, $n = 10$, $n = 20$, and $n = 30$.

   You may stop generating random samples once the distribution of the sample mean is approximately normal.

   c. Record the mean and the standard deviation of the distribution of sample means for each sample size.

	$n = 2$	$n = 5$	$n = 10$	$n = 20$	$n = 30$
mean					
standard deviation					

Sampling from Normal and Non-Normal Populations

    d.   For which sample size is the distribution of sample means approximately normal?

    e.   What is the mean of each distribution (regardless of shape)?

    f.   What is the standard deviation of each distribution (regardless of shape)?

    g.   How does the size of the sample affect the standard deviation of the sample means?

3.   a.   Draw a skewed distribution on the applet by holding the left mouse button down and dragging the mouse over the distribution.

    b.   Obtain 1,000 random samples of size $n = 2$, $n = 5$, $n = 10$, $n = 20$, and $n = 30$.

            You may stop generating random samples once the distribution of the sample mean is approximately normal.

    c.   Record the mean and the standard deviation of the distribution of sample means for each sample size.

	$n = 2$	$n = 5$	$n = 10$	$n = 20$	$n = 30$
mean					
standard deviation					

    d.   For which sample size is the distribution of sample means approximately normal?

    e.   What is the mean of each distribution (regardless of shape)?

    f.   What is the standard deviation of each distribution (regardless of shape)?

    g.   How does the size of the sample affect the standard deviation of the sample means?

Creating a Sampling Distribution for a Proportion

> In this activity, you will randomly select a small group of your classmates and then determine all the possible samples from that population. You will find a proportion for each sample and then construct a sampling distribution for the proportion. The goal of the activity is to better understand the shape, center, and spread of the sampling distribution for the proportion.

This activity will use the data that was collected in the first Chapter 8 activity entitled *Creating a Sampling Distribution for the Mean*. If you did that activity, use the data from it, along with the samples created from it, for the first two parts of this activity.

1. Randomly select six students from the class to treat as a population. Choose a quantitative variable (such as pulse rate, age, or number of siblings) to use for this activity, and gather the data appropriately. Share the data with the class.

    Compute a proportion for this population, as directed by your instructor.

2. a. Work with your group to list all the samples of size $n = 2$, $n = 3$, $n = 4$, or $n = 5$, as assigned by your instructor.

    Collect the results from the other groups to complete the information for all sample sizes.

    Hint: The sample of size $n = 2$ should have $_6C_2 = 15$ possible samples.

    Samples of size $n = 2$ Samples of size $n = 3$

    Samples of size $n = 4$ Samples of size $n = 5$

    b. Compute the sample proportion for each sample listed for your group's sample size in part (a).

    Record the proportions next to the samples.

    Collect the results from the other groups to complete the information for all sample sizes.

98   Creating a Sampling Distribution for a Proportion

c. Form the probability distribution for the sample proportion by making a table for each sample size and listing all the possible sample proportions with their probabilities.

3. Verify for each probability distribution in (2c) that the mean of the sample proportions is the population proportion.

4. Draw the probability histogram for your group on the board. Compare the spread in each probability distribution based on the probability histogram. What does this result imply about the standard deviation of the sample proportion?

Describing the Distribution of the Sample Proportion

> In this activity, you will use an applet to simulate data from a binary distribution. You will describe the center, spread, and shape of the distribution of the sample proportion for various sample sizes.

1. a. Load the **Sampling Distribution Applet** on your computer. Set the applet so that the population is binary with a probability of success equal to 0.25.

   b. Obtain 1,000 random samples of size $n = 5$, $n = 25$, and $n = 75$.

   c. Describe the distribution of the sample proportions based on the results of the applet. That is, describe the center (mean), spread (standard deviation), and shape of the distribution.

	$n = 5$	$n = 25$	$n = 75$
mean			
standard deviation			
shape			

   d. Compare the distributions for the various sample sizes.

   e. What is the mean of the distribution in all three cases (approximately)?

   f. What role does sample size play in the standard deviation?

Chapter 9

Estimating the Value of a Parameter

Exploring the Effects of Confidence Level and Sample Size

> In this activity, you will use an applet to generate confidence intervals for the population proportion. You will use the results to understand what is meant by the confidence level and the role of sample size.

Open the *Confidence Intervals for a Proportion Applet*.

1. Exploring the role of level of confidence:

   a. Construct 1,000 confidence intervals with $n = 100$, $p = 0.3$. What proportion of the 95% confidence intervals included the population proportion, 0.3?

   Construct another 1,000 confidence intervals with $n = 100$, $p = 0.3$. What proportion of the 95% confidence intervals included the population proportion, 0.3?

   Did the same proportion of intervals include the population proportion each time? What proportion did you expect to include the population proportion?

   b. Construct 1,000 confidence intervals with $n = 100$, $p = 0.3$. What proportion of the 99% confidence intervals included the population proportion, 0.3?

   Construct another 1,000 confidence intervals with $n = 100$, $p = 0.3$. What proportion of the 99% confidence intervals included the population proportion, 0.3?

   Did the same proportion of intervals include the population proportion each time? What proportion did you expect to include the population proportion in each part?

2. Exploring the role of sample size:

   a. Construct 1,000 confidence intervals with $n = 10$, $p = 0.3$. What proportion of the 95% confidence intervals included the population proportion, 0.3?

   b. Construct 1,000 confidence intervals with $n = 40$, $p = 0.3$. What proportion of the 95% confidence intervals included the population proportion, 0.3?

Exploring the Effects of Confidence Level and Sample Size

c. Construct 1,000 confidence intervals with $n = 100$, $p = 0.3$. What proportion of the 95% confidence intervals included the population proportion, 0.3?

d. Did the same proportion of intervals include the population proportion in parts (a), (b), and (c)?

What proportion did you expect to include the population proportion in each part?

What happens to the proportion of intervals that include the population proportion as the sample size increases? Why?

Constructing a Confidence Interval with M&M's

> Are the colors evenly distributed in a bag of M&M's? Is the company reporting accurate color percentages for their product? This activity will guide you through constructing a confidence interval for the proportion of M&M's in a particular color. You will compare your confidence interval to those of your classmates and see what percent of them actually contain the proportion reported by the manufacturer.

You have long suspected that the colors are not evenly distributed in a bag of M&M's. Out of curiosity, you decide to construct a confidence interval for the proportion of orange M&M's in a bag of peanut M&M's.

1.   a.   Treat your bag of peanut M&M's as a simple random sample.

       b.   Count the number of orange M&M's and the total number of M&M's in your bag. Then determine the proportion of your M&M's that are orange.

                Number of orange M&M's: _____

                Total number of M&M's: _____

                Proportion of orange M&M's: _____

       c.   Verify that the requirements for constructing a confidence interval for the proportion are satisfied. If the conditions are not met, what should you do?

       d.   Share your data with the rest of the class, according to your instructor's directions.

2.   Construct a 95% confidence interval for the proportion of orange M&M's. Be sure to include an interpretation of your interval.

3.   On the board, construct a scale and draw your confidence interval as a line segment, along with those of your classmates.

106   Constructing a Confidence Interval with M&M's

4.   a.   According to Mars, Inc., peanut M&M's are _____ orange.

       b.   What percent of the class's confidence intervals contained the true proportion of orange M&M's? _____

              What percent of the confidence intervals did you expect to contain the true proportion? Explain.

5.   a.   If you had a larger sample, how would you expect the confidence interval to be affected?

       b.   Pool the class data and assume that all the M&M's constitute one large sample.

              Number of orange M&M's: _____

              Total number of M&M's: _____

              Proportion of orange M&M's: _____

       c.   Construct a 95% confidence interval for the proportion of orange M&M's based on the pooled class data. Be sure to include an interpretation of your interval.

       d.   Does this confidence interval contain the true proportion of orange M&M's?

       e.   How does the margin of error for this confidence interval compare to the margin of error for your individual confidence interval?

Exploring the Effects of Confidence Level, Sample Size, and Shape

In this activity, you will use an applet to generate confidence intervals for the population mean. You will use the results to understand what is meant by the confidence level and the role of sample size. You will also explore how the shape of the original distribution affects the confidence intervals.

Open the **Confidence Intervals for a Mean Applet**.

1. Exploring the role of level of confidence:

    a. Set the shape to normal with mean = 50 and standard deviation = 10.

    b. Simulate at least 1,000 simple random samples from this population with $n = 10$.
    For 95% confidence, what proportion of the confidence intervals included the population mean?

    c. Simulate another 1,000 simple random samples from this population with $n = 10$.
    For 95% confidence, what proportion of the confidence intervals included the population mean?

    d. Did the same proportion of intervals include the population mean in parts (b) and (c)?

    Explain what this result implies.

    What proportion did you expect to include the population mean? Explain any discrepancy.

2. Exploring the role of sample size:

    a. Set the shape to normal with mean = 50 and standard deviation = 10.

    b. Simulate at least 1,000 simple random samples from this population with $n = 10$.
    For 95% confidence, what proportion of the confidence intervals included the population mean?

    c. Simulate 1,000 simple random samples from this population with $n = 20$.
    For 95% confidence, what proportion of the confidence intervals included the population mean?

108   Exploring the Effects of Confidence Level, Sample Size, and Shape

    d.   Simulate 1,000 simple random samples from this population with $n = 40$.
For 95% confidence, what proportion of the confidence intervals included the population mean?

    e.   Compare the width of the intervals for parts (b) – (d).

        Which intervals have the larger width? Explain your response.

3.   Exploring the role of shape:

    a.   Set the shape to skewed right with $n = 5$, mean = 50, and standard deviation = 10.

        Simulate at least 1,000 simple random samples from this population.
For 95% confidence, what proportion of the confidence intervals included the population mean?

    b.   Set the shape to skewed right with $n = 10$, mean = 50, and standard deviation = 10.

        Simulate at least 1,000 simple random samples from this population.
For 95% confidence, what proportion of the confidence intervals included the population mean?

        What proportion of intervals would you expect to include the population mean?

        Is the proportion of simulated intervals that included the population mean closer to what is expected than what was observed in part (a)? Explain.

    c.   Simulate obtaining random samples with a variety of sample sizes. In your judgment, what sample size is needed in order for the proportion of intervals that include the mean to be equal to the proportion suggested by the level of confidence?

        Be sure to clearly describe your approach that led to your conclusion.

Constructing a Confidence Interval from a Non-Normal Distribution

> The exponential probability distribution is a probability distribution that can be used to model waiting time in line or the lifetime of electronic components. Its density function is given by the formula
> $f(x) = \frac{1}{\mu} e^{-x/\mu}$ where $x \geq 0$. Its shape is skewed right. Suppose the wait-time in a line can be modeled by the exponential distribution with $\mu = 5$ minutes.

1. Simulate obtaining a random sample of 200 individuals waiting in line where the mean wait-time is 5 minutes.

   Draw a histogram of the data and comment on its shape.

2. Use StatCrunch, Minitab, or some other statistical software to simulate obtaining 100 simple random samples of size $n = 10$ from the population of 200 individuals waiting in line.

3. a. Determine the mean and standard deviation of the first simulated sample.

      Mean = _____
      Standard deviation = _____

   b. By hand, construct a 95% confidence interval for the mean wait time based on the first sample. Does your interval include the mean of 5 minutes?

110 Constructing a Confidence Interval from a Non-Normal Distribution

4.
   a. Determine 90% *t*-intervals for each of the 100 samples using the statistical software.

   b. How many of the intervals do you expect to include the population mean?

   c. How many of the intervals actually do include the population mean?

   d. Explain what your results suggest.

5. Repeat #2 – 4 for random samples of size $n = 20$ and $n = 30$.

   Write a report detailing the impact of sampling from a non-normal distribution on the actual proportion of intervals that include the population mean versus the expected proportion.

Constructing a Confidence Interval for Die Rolls

How likely are you to roll a 1 with a fair die? How many rolls, on average, should it take to roll a 1? This activity will help you use confidence intervals to explore this question.

1. Either as an individual or in a small group, roll a die until a "1" is observed. Repeat this process a total of 30 times and record the number of rolls required to get a "1" in each of the 30 trials of the experiment.

Trial #	Number of rolls to get a 1	Trial #	Number of rolls to get a 1	Trial #	Number of rolls to get a 1
1		11		21	
2		12		22	
3		13		23	
4		14		24	
5		15		25	
6		16		26	
7		17		27	
8		18		28	
9		19		29	
10		20		30	

2. Use your results to obtain a point estimate of the mean number of rolls of a die required to obtain a "1."

3. Draw a histogram of your data. What is the shape of the histogram?

4. Construct a 90% confidence interval for the mean number of rolls required to obtain a "1."

5. The population mean number of rolls to obtain a "1" is 6. Does your interval include 6? _____

   What proportion of the intervals in the class included 6? _____

   What proportion of the intervals in the class did you expect to include 6? _____

Constructing a Confidence Interval for Die Rolls

Finding a Bootstrap Confidence Interval

> Bootstrapping is sampling with replacement from a sample that is representative of the population whose parameters we wish to estimate. You will obtain many random samples with replacement from the sample data and compute the mean of each random sample. For a 95% confidence interval, you will find the cutoff points for the middle 95% of the sample means. That is, you will find the 2.5th and 97.5th percentiles. These cutoff points represent the lower and upper bounds of the confidence interval.

The data in the table represent survival time (in years) of a random sample of eight patients with a rare disease.

1.0	0.7	0.6	0.5
0.3	1.4	3.1	0.2

1. Verify that 3.1 is an outlier. Use this result to explain why a 95% confidence interval for the mean survival time using Student's t-distribution cannot be determined.

2. Open an Excel spreadsheet and enter the sample data in rows 1 and 2, columns A through D.

   a. To sample with replacement, enter the following command in cell A4:

   =index(a1:d2,rows(a1:d2)*rand()+1,columns(a1:d2)*rand()+1)

   Copy the contents in cell A4 to cells B4 through H4 (because the size of the original sample is $n = 8$, each of our resamples will also have $n = 8$ observations). This creates the first random sample with replacement of size $n = 8$. List the observations in the first resample below.

   b. Copy the contents in cell A4 through H4 to cells A5:H5 through A203:H203. You now have 200 random samples with replacement of $n = 8$. List the observations in the second and third resample below.

   c. Determine the mean of each of the 200 random samples by entering

   = average (A4:H4)

   into cell J4. Copy cell J4 into cells J5 through J203. You now have 200 sample means. List the first three sample means below.

114 Finding a Bootstrap Confidence Interval

3. Draw a histogram of the 200 sample means. Do you believe the distribution of the sample mean is approximately normal? Explain.

4. We want to find the middle $(1-\alpha)100\%$ of the data.

   a. Enter the level of confidence (as a decimal) into cell L1.

   b. In cell L2, enter =1 – L1 to get the value of α. Then enter the following into cell L3 to determine the number of sample means:

   = count(J4:J203)

   c. Find the lower bound by entering the following into cell L4:

   =small(J4:J203,L2/2*L3)

   What is the lower bound?

   d. Find the upper bound by entering the following into cell L5:

   =small(J4:J203,(1-L2/2)*L3)

   What is the upper bound?

   e. Interpret the confidence interval.

5. Move the cursor to an empty cell. Press the "delete" key. What do you notice?

   Use this result to generate five more confidence intervals using the bootstrap.

   Write your intervals below.

Chapter 10

Hypothesis Tests Regarding a Parameter

Interpreting *P*-Values

> A *P*-value is the probability of observing a sample statistic as extreme or more extreme than the one observed under the assumption that the null hypothesis is true. What does this mean in practice though? Can you interpret and understand a *P*-value in the real world? This activity will show you one place that you might see *P*-values in the future and will give you the opportunity to try to interpret them.

Find a prescription drug insert either from your local pharmacy or by browsing online.

Find at least two *P*-values in the prescription drug information and cut or print out the section that references the *P*-value. Attach this here.

Write a few sentences to interpret each *P*-value in the proper context.

120 Testing a Claim with Skittles I

Compare your results to the others in class. Did everyone arrive at the same conclusion for the same color? What about different colors?

Understanding Type I Error Rates I

This activity will utilize an applet to help you understand what it means to make a Type I error and how sample size effects the probability of this type of error.

1. Load the **Hypothesis Tests for a Proportion Applet**. Let the sample size be $n = 200$, True $p = 0.3$, Null $p = 0.3$, and Alternative set to <. Simulate 1000 simple random samples of size $n = 200$.

    a. What would it mean to make a Type I error for this hypothesis test?

    b. Using the $\alpha = 0.05$ level of significance, what proportion of the samples would you expect to lead to a Type I error? What proportion of the samples actually led to a Type I error?

2. Let the sample size be $n = 400$, True $p = 0.3$, Null $p = 0.3$, and Alternative set to <. Simulate 1000 simple random samples of size $n = 400$.

    a. Using the $\alpha = 0.05$ level of significance, what proportion of the samples would you expect to lead to a Type I error? What proportion of the samples actually led to a Type I error?

    b. Compare your results from #1 and #2. Does sample size play a role in the probability of making a Type I error? Explain.

3. Let the sample size $n = 50$, True $p = 0.1$, Null $p = 0.1$, and Alternative set to >. Simulate 1000 simple random samples of size $n = 50$.

    a. What would it mean to make a Type I error for this hypothesis test?

Understanding Type I Error Rates I

b. Using the $\alpha = 0.05$ level of significance, what proportion of the samples would you expect to lead to a Type I error? What proportion of the samples actually led to a Type I error?

c. Repeat the simulation for a sample of size $n = 80$. What proportion of the samples actually led to a Type I error?

d. Repeat the simulation for a sample of size $n = 120$. What proportion of the samples actually led to a Type I error?

e. Explain what happens to the probability of making a Type I error as the sample size increases.

Testing Cola Preferences

> Do you think Coke or Pepsi is more popular? Or do you think they are preferred in equal numbers? This activity will use a simulation to explore and test the assumption that there is no difference in the proportion of males who prefer Coke versus Pepsi.

1. In a survey of 20 males conducted at surveycentral.org, 13 indicated they prefer Coke and 7 indicated they prefer Pepsi.

    a. What proportion of the respondents preferred Coke?

    b. Assuming there is no difference in the proportion of males who prefer Coke over Pepsi, how many of the 20 males in the survey would we expect to prefer Coke?

2. Suppose that reality is that there is no difference in the proportion of males that prefer Coke over Pepsi. If this is the case, we would expect half the respondents to choose Coke, and half to choose Pepsi. Is it possible that this is the case, and we just happened to obtain survey results in which 13 out of 20 prefer Coke? Sure, it is possible, but how possible? To get a sense of the likelihood of obtaining 13 out of 20 males that prefer Coke, we will simulate randomly choosing 20 males from a population in which half prefer Coke and half prefer Pepsi. Essentially, this would be like flipping a coin 20 times and recording the number of heads (for Coke) and the number of tails (for Pepsi).

    a. Flip a coin 20 times and record the number of heads (the number of Coke drinkers) and tails (number of Pepsi drinkers).

Number of heads (Number who prefer Coke)	Number of tails (Number who prefer Pepsi)

    b. Repeat this experiment two more times so that you have three trials with 20 coin flips each.

Number of heads (Number who prefer Coke)	Number of tails (Number who prefer Pepsi)

Copyright © 2013 Pearson Education, Inc.

Testing a Claim with Skittles I

> Skittles Brand candies come in red, orange, yellow, green, and purple (violet). It is well known that M&M Brand candies are not represented with equal frequency. However, do the colors of Skittles candies occur with equal frequency?

1. Write the appropriate null and alternative hypotheses to determine if the proportion of each color is the same in a bag of Skittles.

    H_0:   versus H_1:

2. Randomly choose a color from one of the five colors. Count the total number of candies in the bag as well as the number for the color you selected. What is the sample proportion for the color you chose?

    Number of Skittles in the bag: _____

    Number of Skittles in your color: _____

    Sample proportion for your color: _____

3. Treat the candies in the bag as a simple random sample of all candies produced. Verify the model requirements to conduct an appropriate hypothesis test.

4. Test the hypothesis you wrote in step #1. Be sure to state a conclusion in a complete sentence.

124   Testing Cola Preferences

    c.   Compile the data from your entire class. Draw a dot plot that shows the number of heads (number who prefer Coke) for each trial conducted.

    d.   What proportion of the trials resulted in 13 or more heads (13 or more individuals who prefer Coke)?

3.   To get a true sense as to the likelihood of 13 out of 20 males choosing Coke even if there really is no cola preference, we need to simulate the result thousands of times. Rather than using a coin to perform this simulation, we can use an applet.

    a.   Open the **Coin Tossing Applet**. Set probability of heads to 0.5, number of tosses to 20, and number of repetitions to 1000, and as extreme as to ≥ 13.

    b.   Press the "simulate" button and determine the proportion of repetitions for which we observed 13 or more heads. The result of your simulation represents an approximate *P*-value for the hypothesis test.

    c.   What is the result you obtained? What does this result imply? Does this result suggest that more males prefer Coke over Pepsi?

4.   Suppose the survey resulted in 16 out of 20 males choosing Coke over Pepsi. How would your conclusions change?

5.   Use the binomial probability formula to compute the probability of obtaining 13 or more successes in 20 trials of an experiment for which the probability of success is 0.5. How does this result compare with the simulation results?

Analyzing a New Math Program

> Professors Honey Kirk and Diane Lerma of Palo Alto College developed a "learning community curriculum that blended the developmental mathematics and the reading curriculum with a structured emphasis on study skills." In a typical developmental mathematics course at Palo Alto College, 50% of the students complete the course with a letter grade of A, B, or C. In the experimental course, of the 16 students enrolled, 11 completed the course with a letter grade of A, B, or C.
>
> In this activity you will use simulation to determine the likelihood that 11 or more students out of 16 pass if the course really is not any more effective than the typical developmental math course.
>
> *Source:* Kirk, Honey & Lerma, Diane, "Reading Your Way to Success in Mathematics: A Paired Course of Developmental Mathematics and Reading" *MathAMATYC Educator*; Vol. 1. No. 2 Feb, 2010.

1. What proportion of the students enrolled in the experimental course passed with an A, B, or C?

2. Describe how a coin might be used to simulate the outcome of this experiment to gauge whether the results are unusual.

3. Use MINITAB, StatCrunch, or some other statistical spreadsheet to simulate 1000 repetitions of this experiment assuming the probability a randomly selected student passes the course is 0.5. Or, use the histogram below, which represents the results of 1000 repetitions of the experiment. Use your results or the results below to gauge the likelihood of 11 or more students passing the course if the true pass rate is 0.5. That is, determine the approximate P-value. What does this tell you?

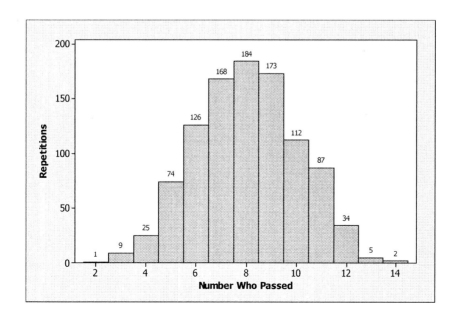

126   Analyzing a New Math Program

4. Use the binomial probability distribution to determine an exact P-value.

5. Now suppose that the actual study was conducted on 48 students and 33 passed the course with an A, B, or C. This would be a study that has ten times as many subjects. What is the proportion of students who passed in this experiment? How does the result compare with #1?

6. Use the **Coin Tossing Applet** to simulate 1000 repetitions of the experiment with 48 students assuming the proportion of students who pass is 0.5. Determine the proportion of repetitions that result in 33 or more passing. That is, determine the approximate P-value.

7. Verify that the normal model can be used to determine the P-value. Then find the P-value. Compare the result with the approximate P-value from #6.

9. In parts (3) and (4) of the activity, we discovered that the P-value was not small enough to reject the null hypothesis that the proportion of students passing the class is 0.5, $H_o: p = 0.5$. *Failing to reject* the statement in the null hypothesis does not mean that we are *accepting* that the proportion is 0.5. Suppose that the researchers, Kirk and Lerma, were told by the Director of Institutional Research that the historical proportion of students who obtain a grade of A, B, or C in developmental mathematics is 0.6. Based on this information, what would be the null and alternative hypothesis?

10. Simulate 1000 repetitions of the experiment where 16 students enroll in the experimental class with each student having a 0.6 probability of passing. Based on the simulation, what is the approximate P-value? That is, what is the likelihood of obtaining 11 or more students who pass the experimental course *if* the true proportion of students who pass is 0.6?

11. Would you reject the statement in the null based on the results obtained in part (b)? Look back at the results obtained in parts (c) and (d). Did you reject the statement in the null? Write a paragraph explaining the difference between *do not reject the statement in the null* and *accept the statement in the null*.

Testing a Claim with Skittles II

> Skittles Brand candies can be purchased in 1-pound bags. But do you really get exactly one pound of Skittles? This activity will help you test that claim from the manufacturer.

1. Write the appropriate null and alternative hypotheses to determine if the mean weight of a bag of Skittles is one pound.

    H_0:   versus H_1:

2. Weigh your bag of Skittles as accurately as possible.

    Exact weight of your bag of Skittles: _____

3. Treat all the bags of Skittles in the class as a simple random sample of all such bags produced. Verify the model requirements to conduct an appropriate hypothesis test.

4. Record the weight of each bag in the class.

5. Test the hypothesis you wrote in step #1. Be sure to state a conclusion in a complete sentence.

128 Testing a Claim with Skittles II

Understanding Type I Error Rates II

This activity will utilize an applet to help you understand what it means to make a Type I error when a hypothesis is conducted on a mean. You will explore how sample size and violating assumptions effect the probability of this type of error.

1. Load the **Hypothesis Tests for a Mean Applet**. Set the shape to normal, the mean to 100, and the standard deviation to 15. These parameters describe the distribution of IQ scores.

    a. Obtain 1000 simple random samples of size $n = 10$ from this population, and test whether the mean is different from 100.

    b. How many of the samples would you expect to lead to a rejection of the null hypothesis if the level of significance is $\alpha = 0.05$?

    c. How many of the 1000 samples led to rejection of the null hypothesis?

2. Set the shape to normal, the mean to 100, and the standard deviation to 15. These parameters describe the distribution of IQ scores.

    a. Obtain 1000 simple random samples of size $n = 30$ from this population, and test whether the mean is different from 100.

    b. How many of the samples would you expect to lead to a rejection of the null hypothesis if the level of significance is $\alpha = 0.05$?

    c. How many of the 1000 samples led to rejection of the null hypothesis?

3. Compare the results of #1 and #2. Does the sample size have any effect on the number of samples that incorrectly rejected the null hypothesis?

4. Set the shape of the distribution to right skewed, the mean to 50, and the standard deviation to 10.

    a. Obtain 3000 simple random samples of size $n = 6$ from this population. Test whether the mean is different from 50.

130　Understanding Type I Error Rates II

  b. How many of the samples led to rejection of the null hypothesis using a level of significance of $\alpha = 0.05$? How many would you expect to lead to rejection of the null hypothesis?

5. Set the shape of the distribution to right skewed, the mean to 50, and the standard deviation to 10.

  a. Obtain 1000 simple random samples of size $n = 60$ from this population. Test whether the mean is different from 50.

  b. How many of the samples led to rejection of the null hypothesis using a level of significance of $\alpha = 0.05$? How many would you expect to lead to rejection of the null hypothesis?

6. Explain why the results of #4 and #5 differ.

Using Bootstrapping to Test a Claim

In Chapter 9, you learned how to use bootstrapping to construct a confidence interval about a parameter. We can also use bootstrapping to test a hypothesis about a parameter. In this activity, you will use bootstrapping to test a hypothesis about a population mean.

The Environmental Protection Agency (EPA) reports that the combined (highway and city driving) miles per gallon (mpg) of a 2010 Ford Fusion Hybrid with a 2.5 liter engine and automatic transmission is 39 mpg. The website fueleconomy.gov allows individuals to report the combined miles per gallon that they are obtaining. The following data represents the miles per gallon of 15 individuals. We want to know whether the data suggest the combined miles per gallon is less than the suggested mpg of 39.

34.1	36.7	36.7	34.9	30.3
37.2	37.2	40.0	36.7	41.0
35.0	39.7	35.6	42.0	39.8

Source: www.fueleconomy.gov

The hypotheses we are testing are

$$H_0: \mu = 39 \text{ versus } H_1: \mu < 39$$

The sample mean of the data is $\bar{x} = 37.13$ mpg and the sample standard deviation is $s = 3.03$ mpg.

We compute the test statistic as

$$t_0 = \frac{37.13 - 39}{3.03/\sqrt{15}} = -2.390$$

The challenge in using the bootstrap to test the hypothesis is that the procedure requires us to assume the data come from a population whose mean is stated in the null hypothesis. Therefore, we need to adjust our data so that it has a mean of 39. This can be accomplished by subtracting the sample mean from the data and then adding the mean stated in the null. For our data, this would mean computing

$$\tilde{x}_i = MPG_i - \overline{MPG} + 39$$

for $i = 1, 2, \ldots, 15$. The transformed data is shown in the table below.

35.97	38.57	38.57	36.77	32.17
39.07	39.07	41.87	38.57	42.87
36.87	41.57	37.47	43.87	41.67

If we compute the mean of the data in the table above, we obtain a value of 39 mpg, which is what we want. So, we now obtain 1000 bootstrap samples from this data. The first three bootstrap samples are shown in the table below.

37.550	38.543	40.043	38.803	39.097	39.683	39.550	39.437	39.757	38.330	39.117	39.163	37.470	39.403	38.757
40.010	38.577	38.777	38.457	39.330	39.963	38.837	39.290	39.077	39.103	38.710	38.857	37.330	39.117	37.423
38.143	39.097	39.110	38.503	40.763	39.503	39.590	39.163	39.297	39.243	38.650	39.670	39.757	39.623	40.063

Using Bootstrapping to Test a Claim

We compute the sample mean, $\bar{\tilde{x}}$, and sample standard deviation, $s_{\tilde{x}}$, which are used to obtain the test statistic, $t_o = \dfrac{\bar{\tilde{x}} - 39}{\dfrac{s_{\tilde{x}}}{\sqrt{15}}}$, for each bootstrap sample. After arranging the test statistics in ascending order, we determine the proportion that are less than $t_o = \dfrac{37.13 - 39}{\dfrac{3.03}{\sqrt{15}}} = -2.390$, which represents the approximation of the P-value.

From the MINITAB output below, we see that 13 of the bootstrap samples result in a test statistic less than -2.390. Our approximate P-value is $13/1000 = 0.013$.

	C1	C2	C3	C4	C5	C6	C7	C8
	Transformed MPG	Bootstrap	Sample	ByVar1	Mean1	StDev1	t	ordered t
1	35.97	38.57	1	1	39.6700	2.21392	1.17208	-5.23784
2	39.07	41.57	1	2	38.6567	3.56127	-0.37338	-4.92661
3	36.87	38.57	1	3	38.3367	3.86387	-0.66490	-3.91153
4	38.57	41.67	1	4	38.6433	2.73926	-0.50428	-3.63501
5	39.07	42.87	1	5	39.8367	2.30517	1.40571	-3.16964
6	41.57	42.87	1	6	39.7300	3.73436	0.75710	-2.83561
7	38.57	39.07	1	7	39.9967	2.30913	1.67165	-2.74463
8	41.87	41.87	1	8	37.8700	2.95876	-1.47916	-2.49048
9	37.47	36.77	1	9	39.4700	2.00036	0.90999	-2.47675
10	36.77	37.47	1	10	39.7433	2.26352	1.27188	-2.46000
11	38.57	41.87	1	11	39.0033	3.36254	0.00384	-2.43356
12	43.87	38.57	1	12	39.1767	3.41373	0.20043	-2.41589
13	32.17	39.07	1	13	38.8900	2.49434	-0.17080	-2.39437
14	42.87	36.77	1	14	38.9700	2.09523	-0.05545	-2.26424
15	41.67	37.47	1	15	38.8300	3.20509	-0.20543	-2.25058
16		38.57	2	16	39.3567	3.28478	0.42053	-2.19008
17		35.97	2	17	39.2367	2.91172	0.31480	-2.18749
18		38.57	2	18	37.7633	3.28557	-1.45777	-2.16002

Using a one-sample t-test to test $H_o : \mu = 39$ versus $H_1 : \mu < 39$, we find the P-value to be 0.016.

Using Bootstrapping to Test a Claim

Now let's have you try one. The following data represents the miles per gallon that 10 individuals experienced with their 2009 Smart car with a 1.0 Liter engine and automatic transmission. We want to know whether the data suggest the combined miles per gallon is greater than the suggested mpg of 36.

34.0	34.6	39.8	36.6	42.9
41.5	46.3	35.3	32.3	43.8

Source: fueleconomy.gov

1. Determine the appropriate null and alternative hypotheses.

2. Compute the test statistic.

3. Transform the original data so that it has a mean equivalent to the mean stated in the null hypothesis.

4. Obtain 1000 bootstrap samples using a statistical spreadsheet.

5. Determine the mean and standard deviation of each bootstrap sample.

6. Compute the test statistic for each bootstrap sample.

7. After arranging the test statistics in ascending order, determine the approximate P-value by finding the proportion of test statistics greater than the value of the test statistic found in #2.

8. Find the P-value using the single sample t-test.

134   Using Bootstrapping to Test a Claim

9. Explain why the original data must be transformed.

10. Devise an approach to use for using bootstrapping on a two-sided hypothesis test.

Computing the Power of a Test

This applet activity will help you explore how the power of a hypothesis test changes as the population proportion approaches the proportion stated in the null hypothesis.

Load the **Hypothesis Tests for a Proportion Applet**. Set the sample size to $n = 100$, True p to 0.44, Null p to 0.5, and alternative to <.

    a.   Simulate obtaining 2000 simple random samples of size $n = 100$.
What proportion of samples led to correctly rejecting the null hypothesis?
That is, what is the power of the test?

    b.   Compute the power of the test for the simulation presented in part (a) by hand.

    c.   Repeat part (a) for True p of 0.46 and 0.48.

    d.   What happens to the power of the test as the true value of the population proportion approaches the proportion stated in the null hypothesis?

Chapter 11

Inferences on Two Samples

Making an Inference about Two Proportions

> A study was conducted by researchers to determine if application of duct tape is more effective than cryotherapy (liquid nitrogen applied to the wart for 10 seconds every 2 to 3 weeks) in the treatment of common warts.
>
> The researchers randomly divided 51 patients into two groups. The 26 patients in Group 1 had their warts treated by applying duct tape to the wart for 6.5 days and then removing the tape for 12 hours, at which point the cycle was repeated, for a maximum of 2 months. The 25 patients in Group 2 had their warts treated by cryotherapy for a maximum of six treatments.
>
> Once the treatments were complete, it was determined that 22 patients in Group 1 (duct tape) and 15 patients in Group 2 (cryotherapy) had complete resolution of their warts. *Source*: Dean R. Focht III, Carole Spicer, Mary P. Fairchok. "The Efficacy of Duct Tape vs. Cryotherapy in the Treatment of Verruca Vulgaris (the Common Wart)," *Archives of Pediatrics and Adolescent Medicine*, 156(10), 2002.

1. What are the researchers trying to show in this study?

2. Explain why the hypotheses to be tested are $H_o: p_1 = p_2$ versus $H_1: p_1 > p_2$.

3. What proportion of the subjects in each group experienced complete resolution of their warts? What is the difference in sample proportions, $\hat{p}_{duct} - \hat{p}_{cryotherapy}$?

4. What does the statement in the null hypothesis suggest about the effectiveness of the two treatments?

5. We need to determine whether the observed differences are due to random error (and there really is no difference in the treatments), or if the differences are significant (so that one treatment is superior to the other). To answer this question, we will randomly assign a treatment to each of the outcomes. The idea is that under the assumption the null hypothesis is true, it should not matter whether a success is the result of duct tape or cryotherapy.

    a. Take a standard deck of cards, and let the 26 red cards represent the duct tape, and let 25 of the black cards represent cryotherapy. Shuffle the cards and then deal 37 cards. These 37 cards represent the 37 individuals who had complete resolution of their warts.

    b. Record the proportion of red cards (duct tape) that had complete resolution of their warts. Record the proportion of black cards (cryotherapy) that had complete resolution of their warts. Now, determine the difference in the proportions, $\hat{p}_{duct} - \hat{p}_{cryotherapy}$.

       Is this proportion as extreme (or more extreme) than the observed difference in proportions?

140 Making an Inference about Two Proportions

6. a. Combine your results from the card simulation with the rest of the class. What proportion of simulations resulted in a sample difference as extreme or more extreme than those observed?

   b. Would you be willing to conclude that duct tape is superior to cryotherapy with this number of observations?

7. What we really want to know is, "If we conducted this experiment many, many times, how likely is it to obtain a sample difference in proportions as extreme, or more extreme than the proportion I observed if there is no difference in the treatments." Certainly, doing this physical simulation many, many times would be labor intensive. So, we will rely on an applet to conduct the simulation for us.

   a. Open the *Randomization Test for Two Proportions: Warts* applet, which is on the CD that accompanies the text or at StatCrunch. Run a single simulation, carefully watching how the applet conducts the randomization.

   b. What is the difference in the sample proportions?

   c. Press randomize 1 time again. Did you get a different difference in the sample proportions?

   d. Now press randomize 1000 times. The applet draws a histogram of the differences in proportions and records the number of simulations that result in differences in the sample proportions as extreme, or more extreme, than the sample difference found in #3.

   Based on the simulation, what might you conclude?

   Are you willing to conclude duct tape is a more effective treatment in dealing with the removal of warts? Explain your position.

Analyzing Rates of Drug Side Effects

Have you ever read the information insert that comes with your prescription medication? Have you ever wondered if the side effects listed are really more likely to occur with the medication than without the medication? This activity will help you analyze the side effect rate for a drug and determine if there is a statistically significant difference in the side effect rates for the drug and placebo groups.

1. Suppose we want to determine which adverse reactions were more common with the prescription drug sildenafil citrate than the placebo, using a significance level of 0.05.

Adverse Event	Percentage of Patients Reporting Event	
	Sildenafil citrate N=734	Placebo N=725
Headache	16%	4%
Flushing	10%	1%
Dyspepsia	7%	2%
Nasal Congestion	4%	2%
Urinary Tract Infection	3%	2%
Abnormal Vision	3%	0%
Diarrhea	3%	1%
Dizziness	2%	1%
Rash	2%	1%

   a. For which side effect are the conditions for the hypothesis testing procedure **not** met? Explain. Do not do the test on that side effect and leave that line in the table blank.

   b. For each side effect listed in the table, test if the proportion of the sildenafil citrate users who experienced the side effect is higher than the proportion of the placebo group who experienced the side effect.

   Summarize your results in the table below.

Adverse Event	P-value	Significantly higher rate for sildenafil citrate users? (Yes/No)
Headache		
Flushing		
Dyspepsia		
Nasal Congestion		
Urinary Tract Infection		
Abnormal Vision		
Diarrhea		
Dizziness		
Rash		

142 Analyzing Rates of Drug Side Effects

2. Do the same treatment for a different prescription drug of your choice. Write up one test completely and summarize the others in the table at the bottom of the page. You can find this information online or you can ask at a local pharmacy for extra prescription information inserts. Either way you obtain the data, please include the information here, in some kind of table or as a screen capture.

Adverse Event	P-value	Significantly higher rate for prescription drug users? (Yes/No)

Comparing Arm Span and Height

> Most people have an arm span approximately equal to their height. This activity will use your class data to test the claim that the mean arm span is the same as the mean height.

1. Each student in your class should measure the length of their arm span and their height, both as accurately as possible in centimeters. Collect the paired data (arm span, height) and record it in the first two columns. Then calculate the differences in arm span and height and record them in the third column.

Arm span (centimeters)	Height (centimeters)	Difference, d (arm span − height)

144 Comparing Arm Span and Height

2. Write the hypotheses necessary to test if there is a difference in the mean arm span and the mean height. Also state the direction of the test.

3. Conduct the hypothesis test at a significance level of 0.05. State the *P*-value.

4. State your conclusion in a complete sentence.

Using Randomization Test for Independent Means

> Do you think male or female students spend more time on homework? Or do you think there's no difference? How would we test this statistically? This activity will guide you through a randomization test for independent means to analyze this issue.

Professor Sullivan noticed in his Elementary Algebra course that the female students appeared to work longer and more diligently on homework assignments than the male students. To confirm his suspicion, he randomly selected 12 male and 12 female student records in MyMathLab and recorded the amount of time on task for a particular section. The results, in minutes, are below.

MALE

81	104	60	83
79	88	88	78
81	74	72	79

FEMALE

92	77	107	99
123	152	116	75
106	81	70	99

Source: MyMathLab

1. Draw dot plots of each data set using the same scale. Does it appear that males are spending less time on their homework? Explain.

2. Calculate the mean time spent on homework for each gender. Then compute the sample mean difference as "Male minus Female." Based on the means, does it appear that males spend less time on their homework? Explain.

3. Certainly, the dotplot and means suggest that males are not spending as much time on their homework. But, are we willing to judge that, in general, males spend less time on homework? Or, is it possible that each gender spends the same amount of time on homework, and we just happened to select some "less than diligent" males? In other words, are the differences between the two samples statistically significant?

    a. Two help answer this question, let's do the following exercise. Take 24 index cards and write the time for each individual on a single card. Shuffle the 24 cards and distribute 12 to the "male" group and 12 to the "female" group. Determine the mean for the "male" group. Determine the mean for the "female" group. Finally, compute the difference in the means as "Male minus Female."

    b. Collect the mean difference from the rest of your classmates. Draw a dotplot of the mean differences below.

Using Randomization Test for Independent Means

    c.   Does it appear to be the case that the differences are centered around 0? Why does this make sense?

    d.   Are any of the results as extreme, or more extreme, than the sample mean difference obtained in #2? What does this suggest?

4.   Open the **Randomization Test for Two Means Applet**. Notice the applet determines the mean difference. Verify the sample means and mean difference from #2.

    a.   Click Randomize 1 time. Explain what the applet is doing when the randomize button is selected. What is the mean difference?

    b.   Now select Randomize 1000 times. What is the center of the histogram of the 1000 (and one) randomized mean differences? Is this what you would expect? What is the shape of the histogram? Is this what you would expect?

    c.   How many of your simulated differences result in a mean difference at or below the value found in #2? What proportion of the randomized differences result in a mean difference at or below the value found in #2?

    d.   Find the *P*-value for the test using Student's *t*-distribution. How does this result compare to the result from part (c)?

5.   Write a conclusion about the study time put in by males versus females.

Differentiating Between Practical and Statistical Significance

> The Mathematics Department at Metropolitan Community College conducted a study in which 62 students who were enrolled in an online course in Intermediate Algebra were divided into two groups. Each group was administered five unit exams plus a proctored, comprehensive final exam. The students in Group 1 had Exams 1, 3, and 5 administered online and the exams were not proctored. Students were allowed access to notes, the text, and other learning aids. However, Exams 2 and 4 were administered as traditional paper and pencil exams that were proctored in a testing center. Students did not have access to notes, the text, or other learning aids. The students in Group 2 took all five exams online and had access to notes, the text, and other learning aids. None of the five exams was proctored. Both groups were administered a paper and pencil comprehensive final exam that was proctored in a testing center.
> *Source:* Flesch, Michael and Ostler, Elliot, "Analysis of Proctored versus Non-proctored Tests in Online Algebra Courses" *MathAMATYC Educator-* Vol. 2, No. 1 August, 2010.

1. What is the response variable in this study? Is it qualitative or quantitative? What is the explanatory variable in this study? Is it qualitative or quantitative?

2. The data below show the results from Test 2, which was proctored for Group 1, but not proctored for Group 2.

	n	Mean	Standard Deviation
Group 1	30	74.30	12.87
Group 2	32	88.62	22.09

   a. Conduct the appropriate test to determine if a statistically significant difference exists between the exam scores in the two groups.

   b. Comment on the standard deviations between the two groups. Specifically, what do the standard deviations imply?

148 Differentiating Between Practical and Statistical Significance

3. The data below show the results from the final exam, which was proctored for both groups.

	n	Mean	Standard Deviation
Group 1	30	65.70	27.66
Group 2	32	74.84	20.10

a. Conduct the appropriate test to determine if a statistically significant difference exists between the exam scores in the two groups.

b. Discuss the practical significance of the results. In other words, regardless of your results from part (a), do the data suggest there is a difference in the two groups?

c. Suppose each group sample size is doubled. Redo part (a) under this scenario.

d. Discuss the difference between statistical and practical significance, in general. Include in your discussion the role sample size might play in obtaining results that are not statistically significant, yet may have practical significance.

Chapter 12

Inference on Categorical Data

Performing a Goodness-of-Fit Test

> This activity will involve comparing the observed color distribution in a bag of M&M's to the expected color distribution that is advertised by the manufacturer. Do you think the advertised color distribution is accurate? How can you decide?

For this activity, you will need a bag of peanut M&M's in the traditional colors (no holiday or special packs).

1. Open your bag of peanut M&M's and count the number of M&M's.

   Number of M&M's in your bag: _____

2. The company that produces M&M products used to advertise on their website that the color distribution for peanut M&M products is as follows:

   Brown   12%
   Yellow 15%
   Red   12%
   Blue   23%
   Orange 23%
   Green   15%

   Based on this information from the manufacturer, calculate the expected number of M&M's of each color in your bag. You will need to consider the total number of M&M's in your bag, along with the advertised proportions. Record these expected frequencies in the table below.

Color	Expected Frequencies	Observed Frequencies	$\dfrac{(\text{observed} - \text{expected})^2}{\text{expected}}$
Brown			
Yellow			
Red			
Blue			
Orange			
Green			
Total			

3. Now count how many M&M's you have in each color in your bag. Enter these observed frequencies in the table above.

4. Are there any observed frequencies in the table that seem very different from the expected frequency? Explain.

152   Performing a Goodness-of-Fit Test

5. Write the hypotheses needed to test if your bag of peanut M&M's has the same proportions as those advertised by the manufacturer.

6. Calculate the value of the chi-square test statistic. Use the last column in the table to record your work.

7. Write a conclusion for the test based on your value of the test statistic or the P-value.

Testing for Homogeneity of Proportions

Have you ever read the information insert that comes with your prescription medication? Have you ever wondered if the side effects listed are more likely to occur with a higher dose of the medication than a lower dose? This activity will help you analyze the side effect rates for a drug and determine if there is a statistically significant difference in the side effect rates for different doses of the drug.

Suppose the following table lists the side effect rates for a particular drug, administered at two different dosages, and the placebo.

Adverse Event	Drug (300 mg/day) ($n = 987$)	Drug (400 mg/day) ($n = 995$)	Placebo ($n = 992$)
Rash	4.8%	1.8%	0%
Nausea	1.6%	3.6%	0.6%
Agitation	0.6%	3.6%	1.6%
Migraine	0%	3.6%	1.6%

1. Calculate the observed and expected frequencies for each cell in the following table for the first adverse event.

Adverse Event	Drug (300 mg/day)	Drug (400 mg/day)	Placebo
Rash			
No rash			

    a   Verify that the conditions are satisfied for the Chi-Square Test for homogeneity of proportions.

    b.   Is there enough evidence to indicate that the proportion of subjects in each group who experienced rash is different at the 0.05 level of significance?

Copyright © 2013 Pearson Education, Inc.

Testing for Homogeneity of Proportions

2. Calculate the observed and expected frequencies for each cell in the following table for the second adverse event.

Adverse Event	Drug (300 mg/day)	Drug (400 mg/day)	Placebo
Nausea			
No nausea			

    a   Verify that the conditions are satisfied for the Chi-Square Test for homogeneity of proportions.

    b.   Is there enough evidence to indicate that the proportion of subjects in each group who experienced nausea is different at the 0.05 level of significance?

3. Calculate the observed and expected frequencies for each cell in the following table for the third adverse event.

Adverse Event	Drug (300 mg/day)	Drug (400 mg/day)	Placebo
Agitation			
No agitation			

    a   Verify that the conditions are satisfied for the Chi-Square Test for homogeneity of proportions.

b. Is there enough evidence to indicate that the proportion of subjects in each group who experienced agitation is different at the 0.05 level of significance?

4. Calculate the observed and expected frequencies for each cell in the following table for the fourth adverse event.

Adverse Event	Drug (300 mg/day)	Drug (400 mg/day)	Placebo
Migraine			
No migraine			

a Verify that the conditions are satisfied for the Chi-Square Test for homogeneity of proportions.

b. Is there enough evidence to indicate that the proportion of subjects in each group who experienced migraine is different at the 0.05 level of significance?

Chapter 13

Comparing Three or More Means

Designing a Randomized Complete Block Design

> How does design affect the flight of a paper airplane? Explore this concept by comparing the flight distances of three different designs. Since the type of paper could have an effect, use the same type of paper (for example, newspaper, brown shipping paper, typing paper, or other) and block by gender.

1. Create a table to record your flight data based on gender (block) and plane design (treatment).

160   Designing a Randomized Complete Block Design

2. Search the Internet for different designs of paper airplanes and choose the designs you wish to use.

   List the designs here and give them names so you can reference them in the table.

3. Discuss how the model design and other variables might affect flight distance. How can you account for these other variables?

4. Fly each plane the same number of times and record the flight distances in your table.

5. Was there a significant difference in flight distances for the different types of designs? Explain.

6. If you answered yes to the previous question, conduct Tukey's test to determine which differences in flight distance are significant using the familywise error rate of 0.05.

Performing a Two-Way ANOVA

> How does the type of paper affect the flight of a paper airplane? In the first part of this activity, we looked at how the design of the plane affected flight distance while we blocked on gender. Now we also want to know whether the paper type plays a role in paper airplane flight. You will use three difference types of paper (for example, newspaper, brown shipping paper, typing paper) and perform a two-way ANOVA.

1. Determine if there is a significant interaction between paper type and design.

2. Draw an interaction plot of the data to support the results.

162   Performing a Two-Way ANOVA

3. a. If there is not significant interaction, determine if there is significant difference in the means for the three plane designs.

   b. If there is not significant interaction, determine if there is significant difference in the means for the paper types.

4. a. If there is significant difference in the means for the three plane designs, use Tukey's test to determine which pairwise means differ using a familywise error rate of 0.05.

   b. If there is significant difference in the means for the paper types, use Tukey's test to determine which pairwise means differ using a familywise error rate of 0.05.

5. Check the requirements for the model. Are they all met? If not, how will that affect your conclusions?

Chapter 14

Inference on the Least-Squares Regression Model and Multiple Regression

Testing the Significance of a Regression Model

> This activity will revisit the arm length and height data collected from your class in Chapter 4. If you have already determined a linear model to relate forearm length and height, you can now test the significance of that model.

This activity will use the data on forearm length and height from the Chapter 4 activity. It will also use the linear regression model that was created in that activity.

1. State the least-squares regression model that was obtained for the forearm and height data in Chapter 4.

2. Verify that the requirements to perform inference on the least-squares regression model are satisfied.

    a. Use an appropriate technology to construct a plot of the residuals against the explanatory variable (forearm length). Explain what you look for in this graph and what condition it helps you verify.

    b. Use an appropriate technology to construct a normal probability plot. Explain what you look for in the normal probability plot and what condition it helps you verify.

3. Test whether a linear relationship exists between forearm length and height at the 0.05 level of significance.

    a. Write the hypotheses needed for this test.

    b. Compute the value of the test statistic or obtain it from an appropriate technology.

    c. Using either the classical or p-value approach, make a decision about the null hypothesis.

    d. State the conclusion.

Testing the Significance of a Regression Model

4.   a.   Construct a confidence interval for the slope of your regression line.

      b.   Write a sentence to interpret this confidence interval.

Using a Randomization Test for Correlation

As stated at the end of Section 14.1, it is generally recommended that inference on the correlation coefficient be avoided because it is difficult to verify the requirement that the data come from a bivariate normal distribution. Instead, it is recommended that the equivalent hypothesis test of $H_o: \beta_1 = 0$ versus $H_o: \beta_1 \neq 0$ be conducted to determine if there is a significant linear relation between two quantitative variables.

Besides testing the hypotheses $H_o: \beta_1 = 0$ versus $H_o: \beta_1 \neq 0$ using the parametric methods presented in the text, we can also use a *permutation test*. A **permutation test** approximates the distribution of a statistic under the assumption the null hypothesis is true (in this case, the data are not linearly related) by randomly assigning a value of the response variable to a value of the explanatory variable and determining the correlation coefficient. This is repeated many times and the proportion of correlation coefficients that are as extreme as or more extreme than the observed correlation coefficient is determined. This value approximates the *P*-value of the distribution under the assumption the null hypothesis is true. The idea is that if the null hypothesis is true, then there is no relation between the explanatory and response variable, so it would be as if the value of the response variable is randomly assigned to the explanatory variable.

1. Consider the following data, which represents the MRI count (a proxy for brain size) and IQ for 10 randomly selected female students.

MRI Count	IQ
816,932	133
951,545	137
991,305	138
833,868	132
856,472	140
852,244	132
790,619	135
866,662	130
857,782	133
948,066	133

   Source: L. Willerrman, R. Schultz, J. N. Rutledge, and E. Bigler (1991). "In Vivo Brain Size and Intelligence," *Intelligence*, 15, 223-228.

   a. Draw a scatter diagram of the data and determine the correlation coefficient between the MRI Count, the explanatory variable, and IQ, the response variable.

   b. What do the scatter diagram and correlation coefficient imply?

168 Using a Randomization Test for Correlation

2. a. Use the "sample from columns" option of a statistical spreadsheet to obtain 20 random samples without replacement from the IQ column. Determine the correlation coefficient between MRI Count and IQ for each of the 20 samples.

   b. Now determine the proportion of correlation coefficients that are greater than the correlation coefficient in #1. This is an estimate of the P-value for the test $H_o : \rho = 0$ versus $H_1 : \rho > 0$, where ρ is the population correlation coefficient between MRI Count and IQ. Explain why.

3. If your statistical spreadsheet (such as StatCrunch) has a resample command, use it to obtain 1000 permutation samples and obtain the resample statistic. The screen below shows how to obtain the resampled statistic from StatCrunch.

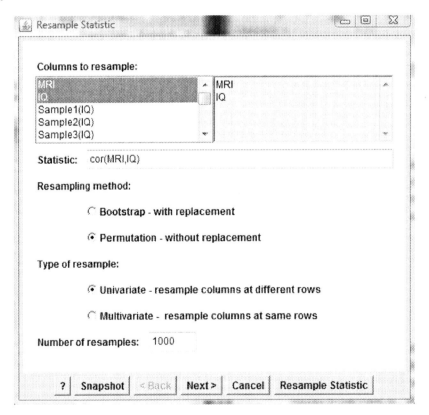

What proportion of the resamples has a correlation coefficient greater than or equal to the observed correlation found in #1?

4. a. Use the **Randomization Test for Correlation Between MRI Count and IQ Applet** to estimate the proportion of resampled correlations that are greater than the correlation found in #1 by running at least 3000 iterations of the randomization applet. That is, use the applet to test $H_o : \rho = 0$ versus $H_1 : \rho > 0$. What is the estimated P-value?

   b. Now conduct the test $H_o : \beta_1 = 0$ versus $H_1 : \beta_1 > 0$. Compare the P-value from this test to the result from part (a).

5. Consider the following data, which represents the estimating selling price (called the "Zestimate") and actual selling price of the home for a random sample of 10 recently sold homes in Oak Park, IL. Both variables are measured in thousands of dollars.

Zestimate	Selling Price
291.5	268
320	305
371.5	360
303.5	283
351.5	340
314	275
332.5	356
295	300
313	285
368	390

   Source: www.zillow.com

   a. Draw a scatter diagram and determine the correlation coefficient between the "Zestimate," the explanatory variable, and selling price, the response variable.

   b. What type of relation appears to exist between the two variables?

Using a Randomization Test for Correlation

c. Use the **Randomization Test for Correlation for Zillow Data Applet** to estimate the proportion of resampled correlations are as extreme or more extreme than the correlation found in part (a) by running at least 3000 iterations of the randomization applet. That is, use the applet to test $H_o: \rho = 0$ versus $H_1: \rho \neq 0$. What is the estimated P-value?

d. Now conduct the test $H_o: \beta_1 = 0$ versus $H_1: \beta_1 > 0$. Compare the P-value from this test to the result from part (c).